CONTROL AND MEASUREMENT OF UNINTENTIONAL ELECTROMAGNETIC RADIATION

CONTROL AND MEASUREMENT OF UNINTENTIONAL ELECTROMAGNETIC RADIATION

W. SCOTT BENNETT

A Wiley-Interscience Publication

JOHN WILEY & SONS, INC.

New York / Chichester / Weinheim / Brisbane / Singapore / Toronto

Library of Congress Cataloging in Publication Data:

Bennett, W. Scott.
    Control and measurement of unintentional electromagnetic radiation
 / W. Scott Bennett.
      p. cm.
    Includes bibliographical references and index.
    ISBN 0-471-17564-1
    1. Shielding (Electricity)   2. Electromagnetic fields-
-Measurement. I. Title.
TK7867.8.B46 1997
621.382′24—dc20   96-34315

CONTENTS

PREFACE

In the past two decades, unintentional electromagnetic radiation from modern electronic systems has become increasingly difficult to control. Digital electronic systems are the primary sources of the problem, and their presence is rapidly becoming an omnipresence. And, a growing recognition of the problem of unintentional electromagnetic radiation has resulted in increasingly strict regulations to control it. However, effective control requires extensive understanding, and digital engineers are seldom extensively schooled in electromagnetic radiation. Furthermore, existing texts and references are directed, for the most part, toward increasing and enhancing electromagnetic radiation, rather than reducing it. Therefore, those texts and references do not provide direct access to the understanding of the problem that is needed by digital designers and other engineers. It is the purpose of this book to fill that need—to provide direct access to basic concepts that must be understood to control unintentional electromagnetic radiation effectively.

The book is organized as follows. Chapter 1 is a brief, general introduction to the primary causes of electromagnetic radiation and to other topics discussed in the remainder of the book. Chapter 2 is a basic examination of intentional electromagnetic radiators, which is included as background for better understanding unintentional radiation. Chapter 3 is a basic examination of the electromagnetic radiation of electrical circuits and how it can be modeled. In Chapter 4, a relatively simple method for decomposing many often-used, periodic voltage waveshapes into their sinusoidal components is developed for use in subsequent chapters. In Chapter 5, useful, easily obtainable descriptions of the measured radiations of periodic, time-varying circuit currents are developed. Chapter 6 contains a discussion of basic methods for controlling the unintentional radiations of electrical circuits based on the models developed and observations made in previous chapters. Chapter 7 is an examination of containment of electromagnetic radiations as a secondary method of control. And, Chapter 8 is an examination of the measurement process with which the effectiveness of controlling unintentional radiations is evaluated.

This book has evolved from seminars presented by the author while a member of the technical staff of Hewlett–Packard Company from 1974 until 1990, and, more recently, as a semi-retired lecturer and consultant. Those seminars and this book were developed for general consumption by anyone with a general engineering background. The book is *not* a detailed catalogue of solutions—it is intended to generate wider understanding and to lead to more effective solutions of a problem that has a definite and growing need for both.

TAILOR THIS BOOK TO YOUR NEEDS

This book consists of generally informative discussions about unintentional electromagnetic radiations and how to control them in the design phase of the manufacture of electronic devices. It is written so that technicians, designers, and managers, or specialists in areas other than electromagnetic compatibility who want basic facts can get them. It is also written for those readers who need a deeper understanding of the underlying physics and the basic mathematics of the problem.

Readers who want basic facts about unintentional electromagnetic radiation and its minimization and control should read Chapter 1; the introductions and summaries of Chapters 2, 3, 4, and 5; and Chapters 6, 7, and 8. Those sections and Chapters 1, 6, 7, and 8 taken alone have been written to form a less mathematically detailed, condensed version of the book.

Those readers who are interested in further background material and foundational discussions should read more of Chapters 2, 3, 4, and 5 than suggested above. And those readers who want to thoroughly absorb its contents should, of course, read the whole book, including the appendices, and any references of interest to them.

Algebra and basic trigonometry are the only mathematics used throughout the book. Other mathematics, when necessary, are either relegated to the appendices, or referenced in the bibliography.

W. SCOTT BENNETT

Loveland, Colorado
September 1996

CHAPTER 1

INTRODUCTION

This book is meant to satisfy the needs of a rather diversified audience. There-fore, if it has not been done already, the preface, "Tailor This Book to Your Needs," should be read by anyone about to join that audience.

1.1 THE PROBLEM

The human environment of today—at work, at home, at school, and at play—con-tains ever-increasing numbers of electronic devices. In recent years it has become increasingly evident that many of these devices interfere with each other and otherwise contaminate the environment. They do so by radiating electromagnetic energy. Some electronic devices are designed to communicate with one another by radiating electromagnetic energy. However, the vast majority of electronic devices are not designed to radiate, yet they very often do radiate. These devices are filling the environment in rapidly increasing numbers and, at the same time, their potential to radiate electromagnetic energy is also rapidly increasing.

Unintentional electromagnetic radiation is a significant engineering problem, as well as a significant societal problem, for many reasons. The reasons can be lumped into three basic categories. (1) Unintentional electromagnetic radia-tion from many electronic devices often interferes with the proper operation of others of those devices. (2) Continued exposure of the human body to relatively low levels of electromagnetic radiation is suspected to be harmful, but its long-term effects are not yet well understood. And, (3) because of (1) and (2), more and more government agencies, worldwide, are imposing limitations on unin-tentional electromagnetic radiations. Thus, the continued design and manufacture of electronic equipment without designing it not to radiate is both impractical and irresponsible.

1.2 THE CAUSES

It is a fundamental physical law that time-varying electric currents radiate electromagnetic energy. And time-varying electric currents are the primary activators of the electronic devices that have become so essential to everyday life. Therefore, the reason that many of those devices radiate electromagnetic energy is obvious. Furthermore, because of another fundamental physical law—the principle of reciprocity—those same radiations interfere with the proper operation of other electronic devices.

The principle of reciprocity says that when the points of excitation and response in a passive, linear, electrical network are interchanged, the excitation-to-response ratio will remain the same. In other words, the transfer of energy is equally efficient in either direction in such a network. This same principle applies, however, if the points of excitation and response are located in separate networks. In other words, reciprocity applies also to separate networks when they are coupled to one another by electromagnetic radiation. Therefore, any electronic device that is a good radiator of electromagnetic energy will also be a good receiver of it. And a good receiver of electromagnetic energy will also be a good radiator of it. The more radiative the time-varying currents of a device are, the more susceptible they will be to corruption—corruption by other currents introduced into their paths by other radiators.

It follows from all of the above that the radiation of time-varying electric currents must somehow be controlled. Otherwise, those currents, and the many useful devices they activate, will have to be eliminated from the environment. Since eliminating time-varying currents or the devices they activate defies the imagination, the former approach is developed here. Principles, rules, and methods, for reducing and controlling the electromagnetic radiations of time-varying currents without interfering with the primary functions of those currents are derived, discussed, and examined. Based on these ideas, it is hoped that, in the very near future, all electronic devices that are not meant to radiate will be carefully and knowingly designed not to radiate. Each of the chapters in this book has been written with that as its primary objective.

1.3 UNDERSTANDING THE PROBLEM

To effectively attack and solve any problem, the first step is to clearly understand it. One good way to do that is to break the problem down into simpler component problems. That is precisely what is done here.

The time-varying currents of most electronic devices in use today are, for the most part, periodic or near-periodic, but they are seldom simple sinusoidal currents. Based on the Fourier theorem, however, a periodically varying waveform can be described as a sum of simple sinusoidal waveforms. Each sinusoidal component in that sum has a frequency that is an integer multiple of the frequency of the given waveform. And each sinusoidal component has an amplitude and phase that are determined by the amplitude and the shape of the original waveform. Thus, any periodic current waveform can be described as a sum of simple sinusoidal current waveforms. That sum is referred to as the *frequency-domain description* of the given current, because each of its sinusoidal components has a different frequency. Although in theory the sum has an infinite number of summands, a small finite number of them will always be sufficient to give an accurate description of the complete sum.

Regulations for controlling unintentional electromagnetic radiations specify that they are to be measured in the frequency domain. In other words, the sinusoidal components of those radiations are measured to verify regulatory compliance. Quite understandably, then, the regulatory limits on unintentional electromagnetic radiations are also specified in the frequency domain. Also, there is a one-to-one relationship between the sinusoidal components of a current and the sinusoidal components of its radiations. Thus, there are clear and definite advantages to working with the sinusoidal components of any current to reduce its radiations.

Based on these observations, the plan of attack here is first to study the sinusoidal radiations of sinusoidal currents that follow relatively simple paths. Those simple sinusoidal radiators are then used to model and understand the radiations of periodic currents that are sums of sinusoidal currents and follow more complicated paths. Armed with the results of those investigations, methods are developed for effectively controlling the radiations of large numbers of often-used time-varying currents.

1.4   THE BASIC RADIATOR

Since some devices radiate and receive electromagnetic energy more effectively than others, there are clearly ways to reduce that radiation and reception, as well as to increase it. Also, since time-varying currents are the causes of electromagnetic radiations, whatever is done to reduce those radiations must be done to the currents that cause them. And, since only *time-varying* currents cause those radiations, two obvious candidates for reducing and controlling the radiations are the speeds with which the currents vary and the amounts they vary. On the other hand, those characteristics of a current clearly don't cause it to receive electromagnetic energy effectively. The only property that could conceivably affect the reception of electromagnetic energy by a current is the geometry of the path it follows. That can be inferred from basic principles of antenna design and from the principle of reciprocity, which also applies to antennas. Thus, radiations of time-varying currents can be reduced by reducing the *speeds* with which they vary, by reducing the *amplitudes* of their variations, and by changing the *paths* that they follow.

These observations can be further verified by examining the mathematical expressions that describe the concept of a basic sinusoidal radiator and its radiated electric field. The *current element* is a short, straight length of sinusoidal current that is assumed to be isolated in empty space. The radiation characteristics of this simple conceptual tool are also simple and easily understood. Therefore, it is used as a basic element with which to analyze and study the radiations of more complicated current configurations.

In mathematical terms, if $i(t) = |I| \sin(2\pi f t)$ is the current of a current element that is centered along the z axis of a three-dimensional coordinate system, it has the radiated electric field

$$E_e(t) = \left(\frac{Z_0}{2cd} \sin \theta \right) L_e |I| f \cos \left[2\pi f \left(t - \frac{d}{c} \right) \right] \qquad (1\text{-}1)$$

In this expression, the factor $(Z_0/2cd)$ consists entirely of constants, which will be defined shortly. As shown in Fig. 1-1, θ is the angle between the z axis and the line from the current element to the observation point of $E_e(t)$. The other variables to which the amplitude of $E_e(t)$ is proportional are $|I|$, the current's amplitude; f, the

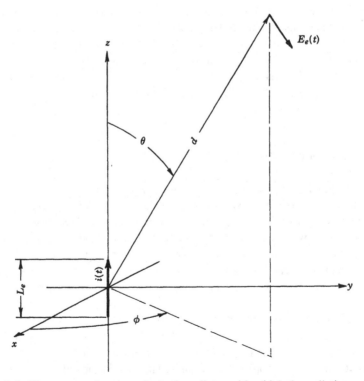

Figure 1-1. The current element—the basic radiator with which the radiations of more complex antenna currents and circuit currents can be found.

current's frequency; and L_e, the length of the current element. In other words, the amplitude of the radiation of a current element depends on the amount the current varies, the speed with which it varies, and the length of its path.

1.5 CONTROLLING THE PROBLEM

The above brief look at the basic radiator reinforces earlier suggestions that the *frequency* with which a current varies, the *amplitude* of its variation, and the *path* it follows are what determine how much it radiates. Those observations will be further reinforced in subsequent discussions. Unfortunately, from a practical point of view, the frequency and amplitude of current variations are largely determined by other design requirements. The path a current follows, however, is almost always relatively arbitrary, so long as it travels a circuit from source, to load, and back. Therefore, without going any further into the details here, it should be obvious that the *path* of a current is the most important design parameter in controlling its radiation.

1.6 CONTAINING THE PROBLEM

The *containment* of electromagnetic radiation, or *shielding* the time-varying currents that cause it from their surroundings, is an idea based on several basic principles of electromagnetics. The first of those principles is that no electromagnetic fields exist in the *interior* of a perfect conductor. There are, of course, no perfect conductors, but *good* conductors, such as copper and silver, are sufficiently close to perfect that little error results if they are assumed to be perfect at the frequencies of interest here.

Another principle, or boundary condition, on which the idea of containment is based is the following. The tangential component of the electric field on either side of the boundary between two different physical media will always be the same. Suppose that boundary is the surface of a near-perfect conductor, then the tangential component of the electric field just outside the surface of that conductor will be zero, as it is in the interior of the conductor.

Another boundary condition basic to containment says that the normal component of the magnetic flux density will always be the same on either side of the boundary between two different physical media. Now, the magnetic flux density is equal to the product of the permeability of a medium and the magnetic field in the medium. Since the permeability of any medium is finite and the magnetic field inside a perfect conductor is zero, it follows that the normal component of the magnetic field will also be zero just outside the surface of a perfect conductor.

Two additional boundary conditions determine what may be the nonzero values of the remaining magnetic and electric field components just outside the surface of a perfect conductor. At the surface of a perfect conductor, (1) the tangential component of the magnetic field is equal in magnitude to the surface current density, and (2) the normal component of the electric field is proportional to the surface charge density. Thus, if the surface of a perfect conductor is well grounded, the surface charge density will be held to zero, by the ground connection. However, charge flow, which is current, is not instantaneous, even in a perfect conductor. Therefore, even perfectly grounded, perfect conductors will sometimes have momentary nonzero surface charges, and the return of that surface charge to zero will create nonzero surface currents. Therefore, the tangential component of the magnetic field and the normal component of the electric field just outside a perfect conductor's surface will not always be zero.

Based on the above boundary conditions, if any time-varying current is placed in a well-grounded, continuous, perfectly conductive container, its radiated fields will definitely be changed and partially contained. However, complete containment is generally out of the question. Also, any attempted containment may interfere with the primary functions a device is meant to perform. These problems and other shortcomings of containment are discussed in greater detail in Chapter 7 of the book.

1.7 MEASURING THE PROBLEM

To determine the extent to which they have been controlled or contained, unintentional electromagnetic radiations are measured according to regulatory specifications. The limits of acceptable radiation are specified, and measurements to verify that those limits are met are made in the frequency domain, as previously noted. The measurement environment specified by regulations is a simulated empty half-space bounded by an infinite, perfectly conducting ground plane. An antenna is positioned above the ground plane to receive radiations from devices being tested, which are also positioned over the ground plane, as illustrated in Fig. 1-2. Antenna responses are fed by cable to a tunable measuring receiver, or spectrum analyzer, which, together with observers, must be located so as to maintain the apparent emptiness of the simulated half-space.

Measured radiations consist of one radiated wave that travels directly to the antenna from the device being tested, and another that arrives there by way of reflection from the ground plane. When the two waves, which are approximately equal in magnitude, arrive at the antenna, depending on their relative phases, they either add to one another or subtract from one another. Therefore, the antenna response is somewhere between zero and twice what it would be if only the direct radiation from the device being tested were received. To standardize these measurements and to determine maximum radiations, the antenna height is varied to find the maximum response at each frequency at which a response occurs. Two different sets of measurements are made—one with the antenna vertically oriented and one with it horizontally oriented. Also, the devices being tested are rotated to find the maximum response with both antenna orientations. The measurements of primary concern here are all made at frequencies of 30 MHz and higher, as specified by various government regulations. If the

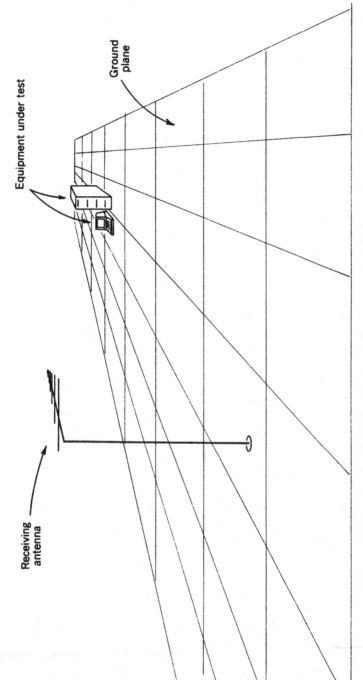

Figure 1-2. The electric fields of unintentional electromagnetic radiations are measured in a simulated empty half-space bounded by a perfectly conducting ground plane.

7

antenna response is below regulatory limits at all frequencies of concern, then any unintentional electromagnetic radiations measured from the devices at those frequencies are considered to be acceptable.

These and other aspects of the measurement of unintentional electromagnetic radiations are discussed in much greater detail in Chapter 8.

1.8 SUMMARY

Unintentional electromagnetic radiation from many electronic devices causes interference with other electronic devices, and over time, electromagnetic radiation may be harmful to the human body. Thus, unintentional electromagnetic radiation is becoming more widely and more strictly regulated, and it must be much more effectively controlled.

Time-varying electric currents radiate electromagnetic energy, and the time-varying currents that activate present-day electronic devices are increasingly liable to radiate increasing amounts of electromagnetic energy. Because of the principle of reciprocity, these devices are also increasingly prone to interference from the radiation of other electronic devices.

To better understand the radiation of any periodic current, the concept of a basic sinusoidal current radiator, the *current element*, is used. Using current elements, more complicated time-varying currents, following more complicated paths, and the resulting radiations can be modeled and more easily understood. And when the causes of those radiations are better understood, they can then be more-easily reduced and controlled.

The amplitudes of electromagnetic radiations are (1) proportional to the frequencies and amplitudes of the time-varying currents that cause them, and (2) highly dependent on the geometries of the paths followed by those currents. Therefore, to control unnecessary electromagnetic radiations, the paths of the currents that cause them must be carefully defined, and the frequencies and amplitudes of the currents must be minimized.

To contain the electromagnetic radiations of time-varying currents, the currents can be placed in a highly conductive, closed container, the outer surface of which is well-grounded to earth. However, even well-grounded, completely closed, perfectly conductive containers will only *partially* contain electromagnetic radiations.

The sinusoidal components of unintentional electromagnetic radiations are measured to verify compliance with government regulations and to verify the effectiveness of the control and/or containment of those radiations. The measurements are performed in a simulated empty half-space over a highly conductive plane that is well grounded to earth. Maximum measured values are sought at each frequency by varying the positions of both the measurement antenna and the devices being tested.

All of the above ideas are examined and discussed in much greater detail in the following chapters.

CHAPTER 2

SINUSOIDAL ANTENNA-CURRENT RADIATIONS

2.1 INTRODUCTION

Electromagnetic radiation is often caused by time-varying currents that are not meant to radiate. The distances currents travel, the geometries of the paths they follow, and the frequencies and amplitudes of their sinusoidal components all contribute to that radiation. Nevertheless, those aspects of current behavior seldom receive sufficient attention during the design and implementation of most electrical and electronic devices. In fact, those properties of currents are seldom precisely known by those who specify their other attributes.

Antenna currents are meant to radiate electromagnetic energy. And to better understand why it radiates, any current or any portion of a current can be viewed as an antenna current. In this book, therefore, an *antenna current* is defined to be any current or any portion of a current that is so viewed to better understand its tendency to radiate.

Clearly, antenna currents should not be included in electronic systems that are not meant to radiate. However, circuit currents are seldom viewed by their designers as antenna currents. *The minimization of antenna-current behavior by circuit currents needs to be an integral part of circuit design.* In other words, the *total* behavior of currents, not only their schematically defined behavior, must be a primary concern of circuit designers. Until that becomes standard design procedure, the effective control of unnecessary electromagnetic radiation will seldom be realized.

It was observed earlier that the electromagnetic radiation of a sinusoidal current is proportional to its amplitude and frequency. In this chapter and the next, the effect on their radiations of the length and geometry of the paths of sinusoidal currents are given primary attention. In later chapters, the combined radiative effects of the lengths and geometries of current paths, and the amplitudes and frequencies of the sinusoidal components of more elaborate currents are studied. In those discussions, *digital circuit currents*, which are sums of sinusoidal currents receive primary attention.

2.2 THE RADIATION OF A CURRENT ELEMENT

The *radiated electric field* of a very short length of sinusoidal current, $i(t) = |I| \sin(2\pi f t)$, can be described as

$$E_e(t) = E_e(\theta) \cos \left[2\pi f \left(t - \frac{d}{c} \right) \right]$$

where

$$E_e(\theta) = \frac{Z_0 |I|}{2d} \frac{f L_e}{c} \sin \theta \qquad (2\text{-}1)$$

This short length of sinusoidally varying current is a *current element*. In these equations, the current element is assumed to lie on the z axis of the coordinate system, with its length centered on the xy plane. Also,

$Z_0 = 120\pi$ ohms, the characteristic impedance of free space
$c = 300 \times 10^8$ meters/second, the speed of light
d = distance, in meters, from the center of the current element to the observation point of $E_e(t)$
$|I|$ = peak amplitude of the current, in amperes
f = frequency of the current, in hertz
L_e = length of the current element, in meters
θ = angle of observation, measured from the current element axis to the direction in which d is measured

By definition, a current segment is a current element only if the current has essentially the same phase over its entire length. Therefore, a current element must be much shorter than one wavelength of its current. A current's wavelength is $\lambda = v_p/f$, where v_p is its phase velocity, and f is its frequency. However, it will very often be the case that $v_p \cong c$, because many dielectrics surrounding current-carrying conductors have permittivities close to that of free space. To simplify the discussion, unless stated otherwise, it is assumed here that $v_p = c$ and current wavelength is $\lambda = c/f$. It is also assumed that any straight-line segment of current of length $L \leq \lambda/16$ can be viewed as a current element.[1]

As seen from Eq. 2-1, the radiated electric field of a current element is directly proportional to the quantity $L_e f/c$. That quantity is equal to the length of a current element expressed as a fraction of a wavelength. Therefore, this quantity will be denoted as $L_{e\lambda}$. In other words, $L_{e\lambda} \equiv L_e f/c$ is the length of a current element in *radiation* wavelengths. With this definition, the length restriction for current elements becomes $L_{e\lambda} \ll 1$. Thus, a current segment is a current element whenever its length is such that $L_{e\lambda} \to 0$. With the definition $L_{e\lambda} \equiv L_e(f/c) \to 0$, the latter of Eq. 2-1 can be written as

$$E_e(\theta) = \frac{Z_0 |I| L_{e\lambda}}{2d} \sin \theta \qquad (2\text{-}2)$$

When a current element is centered along the z axis, as seen from Eqs. 2-1 and 2-2, $E_e(\theta)$, the peak amplitude of $E_e(t)$, is directly proportional to $\sin \theta$ and independent of ϕ. Independence of ϕ, which is measured in the xy plane or in planes parallel to the xy plane, means that $E_e(\theta)$ is symmetric about the z axis, which is its own axis. Thus, from Eqs. 2-1 and 2-2, the radiation pattern of a current element, which is a polar plot of $|E_e(\theta)|$ for a fixed distance $d \gg c/f$, is seen to have the shape of a perfectly round doughnut with a vanishing hole. Cross sections of this radiation pattern are shown in Fig. 2-1, together with a three-dimensional perspective of the pattern.

The distance from the center of a current segment to the surface of its radiation pattern in any direction indicates the relative magnitude of its radiated electric field in that direction. For example, suppose the observation point of the radiated electric field of the current element of Fig. 2-1 is in the xy plane where the angle $\theta = \pi/2$. Since $\sin(\pi/2) = 1$, $\sin \theta \leq 1$, and the other factors of $E_e(\theta)$ are constants, it is clear that $|E_e(\theta)|$ is a maximum in the xy plane. Therefore, the radiated electric field of a current element is a maximum in the plane to which it is normal and passes through its center. When the observation point of $E_e(t)$ is on the axis of the current element, the z axis here, $|E_e(\theta)| = 0$, because on that axis $\theta = 0$, and $\sin \theta = \sin(0) = 0$. Thus, the radiated electric field of a current element is zero along the axis with which it is coaxial.

The radiated electric field of any straight segment of current of length L can be given the same general description as that of a current element. For example, if the current at the center of the current segment shown in Fig. 2-2 is

$$i(t) = |I| \sin(2\pi f t)$$

---

[1]If more accurate approximations of current wavelengths are necessary, they may be obtained by using $v_p = c/\sqrt{k}$, where k is the relative permittivity of the dielectric surrounding the conductor that carries the current.

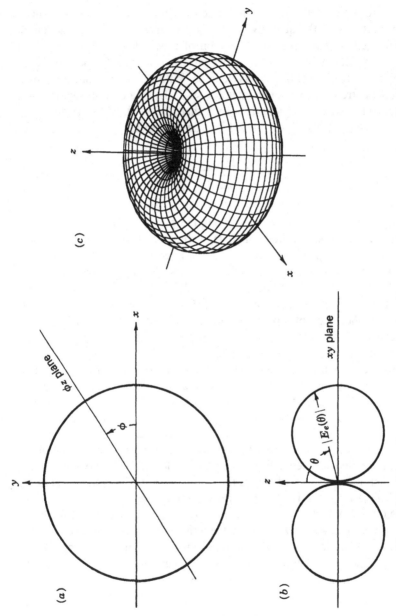

(a)

(b)

(c)

Figure 2-1. The radiation pattern of a current element when it lies on the z axis, centered on the xy plane: (a) its xy plane cross section, (b) its φz plane cross section, and (c) a three-dimensional view.

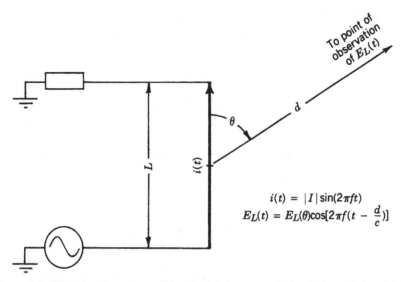

Figure 2-2. The situation assumed to obtain the general description of the radiated electric field $E_L(t)$ of a segment of current $i(t)$ of length L.

then its radiated electric field is

$$E_L(t) = E_L(\theta) \cos \left[2\pi f \left(t - \frac{d}{c} \right) \right] \qquad (2\text{-}3)$$

When θ is the observation angle, the segment is assumed to be aligned with and centered on the z axis, and $E_L(\theta)$ is its *radiation pattern factor*. The distance d is always measured from the center of the segment, and the pattern factor is the peak amplitude of $E_L(t)$ in the direction of θ. Thus, the radiation pattern of $E_L(t)$ is a polar plot of $|E_L(\theta)|$, the magnitude of the pattern factor.

From Eqs. 2-1 and 2-3 it can be seen that the general description of the radiated electric field of any straight-line segment of current is the same as that of the current element. However, few current segments longer than current elements have radiation patterns that are as simply describable as that of the current element. Nevertheless, the pattern factor of any straight-line current segment can be found as shown below. And determining the pattern of a radiation is often the first step in understanding how best to reduce and control it.

2.3 RADIATION PATTERNS OF LONGER CURRENT SEGMENTS

Descriptions of the radiated electric fields of current segments that are longer than current elements can be found as follows. Consider the two current segments shown in Fig. 2-3. Each has length $L \geq L_e$, and at the point where they meet the current is $i(t) = |I| \sin(2\pi f t)$. Because the current phase velocity is $v_p \leq c$ in the upward direction, the current at the center of the upper segment will be

$$i\left(t - \frac{L}{2v_p}\right) = |I| \sin\left[2\pi f\left(t - \frac{L}{2v_p}\right)\right] \qquad (2\text{-}4)$$

At the center of the lower segment, which has the same length, the current will be

$$i\left(t + \frac{L}{2v_p}\right) = |I| \sin\left[2\pi f\left(t + \frac{L}{2v_p}\right)\right] \qquad (2\text{-}5)$$

The distances to the observation point, as shown in Fig. 2-3, will be d from the point where the segments meet, $d - (L/2) \cos \theta$ from the center of the upper segment, and $d + (L/2) \cos \theta$ from the center of the lower segment. Therefore, substituting $d - (L/2) \cos \theta$ for d in Eq. 2-3, the radiated electric field of the

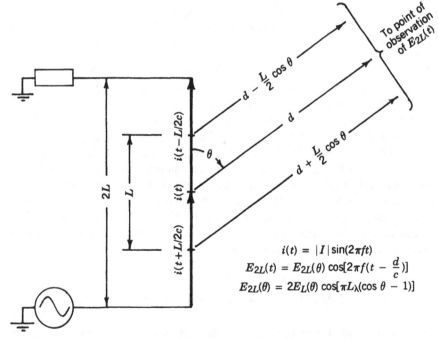

$$i(t) = |I| \sin(2\pi f t)$$
$$E_{2L}(t) = E_{2L}(\theta) \cos[2\pi f(t - \frac{d}{c})]$$
$$E_{2L}(\theta) = 2E_L(\theta) \cos[\pi L_\lambda(\cos \theta - 1)]$$

Figure 2-3. The radiated electric field of two adjacent, in-line current segments, each of length L with the pattern factor $E_L(\theta)$, is equal to $E_{2L}(t)$ with the pattern factor $E_{2L}(\theta)$.

upper segment, when $L_\lambda = Lf/c$ is the segment length in radiation wavelengths, will be

$$E_u(t) = E_u(\theta) \cos \left\{ 2\pi f \left[t - \frac{L}{2v_p} - \frac{(d - (L/2)\cos\theta)}{c} \right] \right\}$$

$$= E_L(\theta) \cos \left[2\pi f \left(t - \frac{d}{c} \right) + \pi L_\lambda \left(\cos\theta - \frac{c}{v_p} \right) \right] \qquad (2\text{-}6a)$$

Or, if $v_p = c$,

$$E_u(t) = E_L(\theta) \cos \left[2\pi f \left(t - \frac{d}{c} \right) + \pi L_\lambda(\cos\theta - 1) \right] \qquad (2\text{-}6b)$$

When $d + (L/2) \cos\theta$ is substituted for d in Eq. 2-3, the radiated electric field of the *lower* segment will be

$$E_\ell(t) = E_\ell(\theta) \cos \left[2\pi f \left(t + \frac{L}{2v_p} - \frac{d + (L/2)\cos\theta}{c} \right) \right]$$

$$= E_L(\theta) \cos \left[2\pi f \left(t - \frac{d}{c} \right) - \pi L_\lambda \left(\cos\theta - \frac{c}{v_p} \right) \right] \qquad (2\text{-}7a)$$

Of, if $v_p = c$,

$$E_\ell(t) = E_L(\theta) \cos \left[2\pi f \left(t - \frac{d}{c} \right) - \pi L_\lambda(\cos\theta - 1) \right] \qquad (2\text{-}7b)$$

The pattern factors will be $E_u(\theta) = E_\ell(\theta) = E_L(\theta)$, because in the radiated electric field $L \ll d$ and $d \pm (L/2) \cos\theta \cong d$. In other words, the pattern factors will be essentially the same, because the observation point is a long distance from the current segments.

As given in Appendix A, $\cos(x + y) + \cos(x - y) = 2 \cos(x) \cos(y)$ for any x and y. Therefore, if $x = 2\pi f(t - d/c)$ and $y = \pi L_\lambda(\cos\theta - 1)$, and assuming that $v_p = c$, it follows from Eqs. 2-6 and 2-7 that

$$E_{2L}(t) = E_u(t) + E_\ell(t) = E_{2L}(\theta) \cos \left[2\pi f \left(t - \frac{d}{c} \right) \right]$$

where

$$E_{2L}(\theta) = 2E_L(\theta)\cos[\pi L_\lambda(\cos\theta - 1)]$$

$$= E_L(\theta)\,\frac{\sin[2\pi L_\lambda(\cos\theta - 1)]}{\sin[\pi L_\lambda(\cos\theta - 1)]} \tag{2-8}$$

The final expression for $E_{2L}(\theta)$ follows from the identity $2\sin(x)\cos(x) = \sin(2x)$, which is also given in Appendix A. Equation 2-8 describes the radiated electric field of a segment of sinusoidal current when it is upwardly propagating over a length $2L_\lambda$.

Example 2-1 The pattern factor $E_{2e}(\theta)$ of a straight-line current segment with a length equal to that of two current elements can be found as follows. The length of such a segment is, of course, $2L_e$ or $2L_{e\lambda}$, where $L_{e\lambda} \to 0$. From Fig. 2-3 and Eqs. 2-2 and 2-8, it can be seen that

$$E_{2e}(t) = E_{2e}(\theta)\cos\left[2\pi f\left(t - \frac{d}{c}\right)\right]$$

And assuming that $v_p = c$,

$$E_{2e}(\theta) = 2[E_e(\theta)]\cos[\pi L_{e\lambda}(\cos\theta - 1)]$$

$$= \left(\frac{Z_0|I|L_{e\lambda}}{2d}\,\sin\theta\right)\frac{\sin[2\pi L_{e\lambda}(\cos\theta - 1)]}{\sin[\pi L_{e\lambda}(\cos\theta - 1)]} \tag{2-9}$$

Since $L_{e\lambda} \to 0$, it follows that

$$\sin[\pi L_{e\lambda}(\cos\theta - 1)] = \pi L_{e\lambda}(\cos\theta - 1)$$

and Eq. 2-9 can be written as

$$E_{2e}(\theta) = \left(\frac{Z_0|I|}{2\pi d}\,\sin\theta\right)\frac{\sin[2L_{e\lambda}\pi(\cos\theta - 1)]}{\cos\theta - 1}$$

$$= \left(\frac{E_e(\theta)}{\pi L_{e\lambda}}\right)\frac{\sin[2L_{e\lambda}\pi(\cos\theta - 1)]}{\cos\theta - 1} \tag{2-10}$$

Cross sections of the radiation patterns of $E_e(\theta)$ and $E_{2e}(\theta)$ for the same current $i(t)$ are illustrated in Fig. 2-4a and b.

Example 2-2 Given the value of $E_{2e}(\theta)$ in Eq. 2-10, $E_{4e}(\theta)$, the radiation pattern factor of a straight-line current segment equal in length to four current elements, can be found by letting $L = 2L_e$ in Eq. 2-8. The component segments

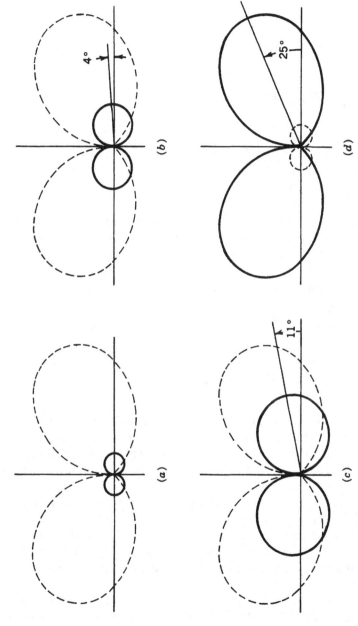

Figure 2-4. Comparisons of cross sections of the radiation patterns of a current element of length $L_e = \lambda/16$, and current segments of length $2L_e$, $4L_e$, and $8L_e$, when the same sinusoidal current is propagating upward in each: (a) $|E_e(\theta)|$; (b) $|E_{2e}(\theta)|$; (c) $|E_{4e}(\theta)|$; (d) $|E_{8e}(\theta)|$.

in Fig. 2-3 will then both have length $L_\lambda = 2L_{e\lambda}$. Therefore, when $v_p = c$, the pattern factor for a current segment of length $4L_{e\lambda}$ will be

$$E_{4e}(\theta) = 2E_{2e}(\theta)\cos[2L_{e\lambda}\pi(\cos\theta - 1)]$$

$$= 2\left\{\frac{E_e(\theta)}{\pi L_{e\lambda}}\ \frac{\sin[2L_{e\lambda}\pi(\cos\theta - 1)]}{\cos\theta - 1}\right\}\cos[2L_{e\lambda}\pi(\cos\theta - 1)]$$

$$= \left[\frac{E_e(\theta)}{\pi L_{e\lambda}}\right]\frac{\sin[4L_{e\lambda}\pi(\cos\theta - 1)]}{\cos\theta - 1} \tag{2-11}$$

The latter expression for $E_{4e}(\theta)$ follows again from the identity $2\sin(x)\cos(x) = \sin(2x)$, given in Appendix A, when it is assumed that $x = 2L_{e\lambda}\pi(\cos\theta - 1)$. The radiation pattern $|E_{4e}(\theta)|$ is illustrated in Fig. 2-4c.

Example 2-3   Next suppose that the component segments in Fig. 2-3 each have length $L_\lambda = 4L_{e\lambda}$. Then their combined length will be $8L_e$, and, when $v_p = c$, the pattern factor will be

$$E_{8e}(\theta) = 2E_{4e}(\theta)\cos[4L_{e\lambda}\pi(\cos\theta - 1)]$$

$$= 2\left\{\frac{E_e(\theta)}{\pi L_{e\lambda}}\ \frac{\sin[4L_{e\lambda}\pi(\cos\theta - 1)]}{\cos\theta - 1}\right\}\cos[4L_{e\lambda}\pi(\cos\theta - 1)]$$

$$= \left[\frac{E_e(\theta)}{\pi L_{e\lambda}}\right]\frac{\sin(8L_{e\lambda}\pi(\cos\theta - 1))}{\cos\theta - 1} \tag{2-12}$$

The latter expression for $E_{8e}(\theta)$ follows again from the identity $2\sin(x)\cos(x) = \sin(2x)$, where, in this case, $x = 4L_{e\lambda}\pi(\cos\theta - 1)$. The radiation pattern $|E_{8e}(\theta)|$ is plotted in Fig. 2-4d for comparison with the radiation patterns $|E_e(\theta)|, |E_{2e}(\theta)|$, and $|E_{4e}(\theta)|$. The current $i(t)$ is assumed to be the same in each case.

Notice in Fig. 2-4 that as the segments get longer, their maximum radiations increase, and the directions of maximum radiation move to higher angles above the planes on which they are centered. Thus, in three dimensions, cones of radiation are forming around the segments as they get longer. It is interesting to see how this trend continues as the radiation patterns of even longer segments are examined.

Now, from Eqs. 2-10, 2-11, and 2-12, it can readily be seen that for any positive integer n, if $N = 2^n$ and $v_p = c$, then

$$E_{Ne}(\theta) = \left[\frac{E_e(\theta)}{\pi L_{e\lambda}}\right]\frac{\sin[NL_{e\lambda}\pi(\cos\theta - 1)]}{\cos\theta - 1} \tag{2-13}$$

Thus, the pattern factor of any current segment of length $L_\lambda = NL_{e\lambda}$, where $N = 2^n$, $0 < L_{e\lambda} \leq \frac{1}{16}$, and $v_p = c$, is given by Eq. 2-13. As a result, since $L_{e\lambda}$ can be made to approach zero, and n can be any positive integer, $NL_{e\lambda}$ can be made equal to any length, L_λ. In other words, the pattern factor of any straight-line current segment, where L_λ is its length in wavelengths, is

$$E_L(\theta) = \left[\frac{E_e(\theta)}{\pi L_{e\lambda}} \right] \frac{\sin[L_\lambda \pi(\cos\theta - 1)]}{\cos\theta - 1}$$

$$= \left(\frac{Z_0|I|}{2\pi d} \sin\theta \right) \frac{\sin[L_\lambda \pi(\cos\theta - 1)]}{\cos\theta - 1} \tag{2-14a}$$

when $v_p = c$, and

$$E_L(\theta) = \left[\frac{E_e(\theta)}{\pi L_{e\lambda}} \right] \frac{\sin[L_\lambda \pi(\cos\theta - c/v_p)]}{\cos\theta - c/v_p}$$

$$= \left(\frac{Z_0|I|}{2\pi d} \sin\theta \right) \frac{\sin[(L_\lambda \pi(\cos\theta - c/v_p)]}{\cos\theta - c/v_p} \tag{2-14b}$$

when $v_p < c$. The radiation patterns of numerous current segments are found in the following examples, to illustrate the wide applicability of Eqs. 2-14. These equations describe the radiation pattern of any straight-line current segment of length L_λ, so long as $L \ll d$.

Example 2-4   First the pattern factors will be found for all current segments of lengths less than $8L_e$ that are integral multiples of $L_{e\lambda}$ and were previously skipped over. These can all be obtained with Eq. 2-14, assuming $L_{e\lambda} = \frac{1}{16}$ and $v_p = c$. Thus, a current segment of length $3L_{e\lambda} = \frac{3}{16}$ will have the radiation pattern factor

$$E_3(\theta) = \left(\frac{Z_0|I|}{2\pi d} \sin\theta \right) \frac{\sin[3\pi(\cos\theta - 1)/16]}{\cos\theta - 1} \tag{2-15}$$

The radiation pattern $|E_3(\theta)|$ is illustrated along with the others of the eight shortest current segments in Fig. 2-5. The remaining pattern factors of those segments, when $L_\lambda = 5L_{e\lambda}$, $6L_{e\lambda}$, and $7L_{e\lambda}$, are

$$E_5(\theta) = \left(\frac{Z_0|I|}{2\pi d} \sin\theta \right) \frac{\sin[5\pi(\cos\theta - 1)/16]}{\cos\theta - 1}$$

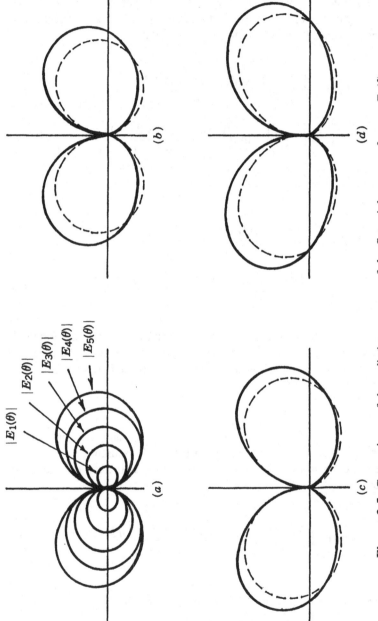

Figure 2-5. Comparisons of the radiation patterns of the first eight pattern factors, $E_n(\theta)$, assuming that $E_1(\theta) = E_e(\theta)$ and $L_e = \lambda/16$, as given in Eq. 2-2: (a) the first five radiation patterns; (B) $|E_6(\theta)|$ compared to $|E_5(\theta)|$; (c) $|E_7(\theta)|$ compared to $|E_6(\theta)|$; (d) $|E_8(\theta)|$ compared to $|E_7(\theta)|$.

$$E_6(\theta) = \left(\frac{Z_0|I|}{2\pi d} \sin\theta \right) \frac{\sin[3\pi(\cos\theta - 1)/8]}{\cos\theta - 1}$$

and

$$E_7(\theta) = \left(\frac{Z_0|I|}{2\pi d} \sin\theta \right) \frac{\sin[7\pi(\cos\theta - 1)/16]}{\cos\theta - 1} \tag{2-16}$$

The radiation patterns obtained with $|E_5(\theta)|$, $|E_6(\theta)|$, and $|E_7(\theta)|$ are also shown in Fig. 2-5.

Example 2-5 In Fig. 2-6, the radiation patterns of several upward-propagating, sinusoidal currents in segments of lengths greater than those considered above are illustrated. The pattern factors for those segments are obtained using Eq. 2-14, with $v_p = c$, except for $E_8(\theta) = E_{8e}(\theta)$, which was given in Eq. 2-12. Assuming that $L_{e\lambda} = \frac{1}{16}$, they are

$$E_{10}(\theta) = \left(\frac{Z_0|I|}{2\pi d} \sin\theta \right) \frac{\sin[5\pi(\cos\theta - 1)/8]}{\cos\theta - 1}$$

$$E_{12}(\theta) = \left(\frac{Z_0|I|}{2\pi d} \sin\theta \right) \frac{\sin[3\pi(\cos\theta - 1)/4]}{\cos\theta - 1}$$

$$E_{14}(\theta) = \left(\frac{Z_0|I|}{2\pi d} \sin\theta \right) \frac{\sin[7\pi(\cos\theta - 1)/8]}{\cos\theta - 1}$$

$$E_{16}(\theta) = \left(\frac{Z_0|I|}{2\pi d} \sin\theta \right) \frac{\sin[\pi(\cos\theta - 1)]}{\cos\theta - 1}$$

$$E_{18}(\theta) = \left(\frac{Z_0|I|}{2\pi d} \sin\theta \right) \frac{\sin[9\pi(\cos\theta - 1)/8]}{\cos\theta - 1}$$

$$E_{20}(\theta) = \left(\frac{Z_0|I|}{2\pi d} \sin\theta \right) \frac{\sin[5\pi(\cos\theta - 1)/4]}{\cos\theta - 1}$$

and

$$E_{22}(\theta) = \left(\frac{Z_0|I|}{2\pi d} \sin\theta \right) \frac{\sin[11\pi(\cos\theta - 1)/8]}{\cos\theta - 1} \tag{2-17}$$

The radiation patterns obtained with these equations are all shown in Fig.

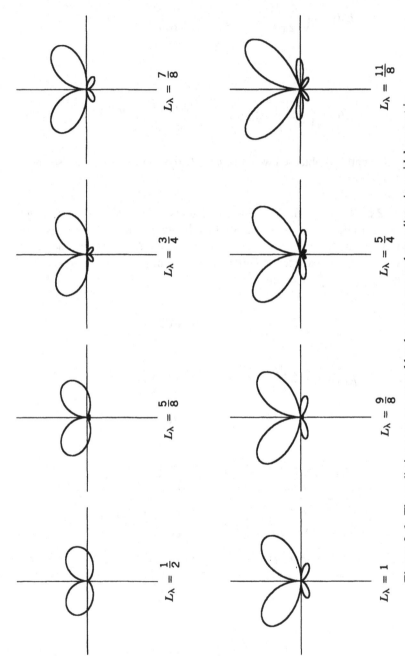

Figure 2-6. The radiation patterns caused by the same upward-traveling sinusoidal current in current segments of the lengths shown.

$L_\lambda = \frac{1}{2}$

$L_\lambda = \frac{5}{8}$

$L_\lambda = \frac{3}{4}$

$L_\lambda = \frac{7}{8}$

$L_\lambda = 1$

$L_\lambda = \frac{9}{8}$

$L_\lambda = \frac{5}{4}$

$L_\lambda = \frac{11}{8}$

2-6, along with that of $|E_8(\theta)|$, for comparison. The same sinusoidal current is assumed to exist in each segment.

The radiation patterns in Fig. 2-6 show that a second cone of radiation becomes part of the pattern when a current segment's length lies between $L_\lambda = 8L_{e\lambda} = \frac{1}{2}$ and $L_\lambda = 16L_{e\lambda} = 1$. A third cone of radiation has begun to form in the pattern $|E_{18}(\theta)|$, for which the segment length is slightly larger than one full wavelength, and becomes quite visible in patterns $|E_{20}(\theta)|$, and $|E_{22}(\theta)|$. In fact, whenever a current segment's length L_λ is such that $n/2 < L_\lambda \le (n+1)/2$, where n is a positive integer, its radiation pattern will have $n+1$ cones of radiation. Further evidence of this can be seen in the following example.

Example 2-6   In Fig. 2-7, the radiation patterns of segments of equal, upward-propagating, sinusoidal currents having lengths of $24L_e$, $26L_e$, $28L_e$, $30L_e$, $32L_e$, $34L_e$, $36L_e$, and $38L_e$ are shown. Pattern factors for these current segments are found using Eq. 2-14, with $L_{e\lambda} = \frac{1}{16}$ and $v_p = c$, as they were in the previous example.

One further observation should be made here. If the sources and loads in Figs. 2-2 and 2-3 were to have their positions interchanged, then the currents in the segments in those figures would propagate downward. That would cause the radiation patterns of those segments to be rotated 180°, or π radians, relative to their axes. To account for that rotation when the segment lies on the z axis, it is necessary only to replace θ with $\theta - \pi$ in the pattern factor of the upward propagating current.

Thus, if the currents in Figs. 2-2 and 2-3 were all propagating downward rather than upward, then Eq. 2-3 would be replaced with

$$E_L(t) = E_L(\theta - \pi)\cos\left[2\pi f\left(t - \frac{d}{c}\right)\right]$$

where

$$E_L(\theta - \pi) = -\left(\frac{Z_0|I|}{2\pi d}\sin\theta\right)\frac{\sin(L_\lambda\pi[\cos\theta + c/v_p])}{\cos\theta + c/v_p} \qquad (2\text{-}18)$$

Also, $E_e(\theta - \pi) = -E_e(\theta)$, because $\sin(\theta - \pi) = -\sin\theta$ replaces $\sin\theta$ in the expression for $E_e(\theta)$.

To summarize the above, a relatively simple mathematical expression describing the radiation pattern of a straight segment of sinusoidal current of any length is easily obtained. With that description the radiation pattern of any such current segment can be viewed graphically. It is seen that both the size and shape of a current segment's radiation pattern vary with its length L_λ, but

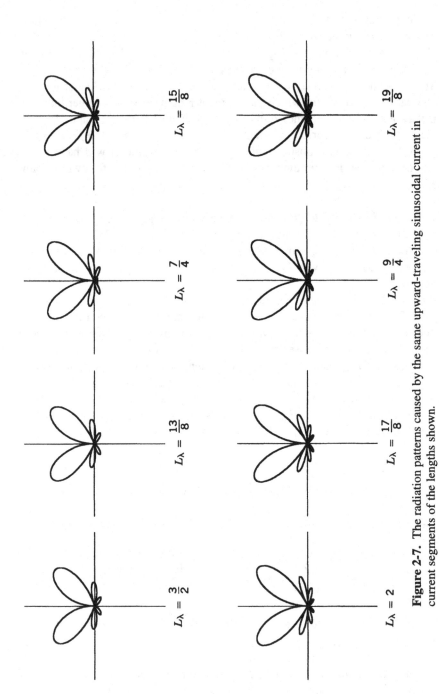

Figure 2-7. The radiation patterns caused by the same upward-traveling sinusoidal current in current segments of the lengths shown.

$L_\lambda = \frac{3}{2}$ $L_\lambda = \frac{13}{8}$ $L_\lambda = \frac{7}{4}$ $L_\lambda = \frac{15}{8}$

$L_\lambda = 2$ $L_\lambda = \frac{17}{8}$ $L_\lambda = \frac{9}{4}$ $L_\lambda = \frac{19}{8}$

the magnitude of the radiation is not directly proportional to L_λ. Additional cones of radiation do form in the radiation pattern of a current segment as its length increases, however. One additional cone is formed with each additional half-wavelength of segment length.

Radiation patterns considered so far have been those of straight current segments with only one current component. The radiation patterns of applied and reflected currents are now examined.

2.4 RADIATIONS OF OPEN-ENDED CURRENT SEGMENTS

In the above discussion, it was assumed that there were no current reflections, but currents are often reflected, and that changes the radiation pattern. More specifically, in Fig. 2-8, suppose the current arriving at the open end of the conductor of length $L_\lambda = Lf/c$ is $i(t) = |I| \sin(2\pi f t)$. Then, assuming $v_p = c$, and taking phase changes into account, a distance $L_\lambda/2$ back from the open end, at the center of the segment, the current propagating *toward* the end will be

$$i_a\left(t + \frac{L}{2c}\right) = |I| \sin(2\pi f t + \pi L_\lambda) \tag{2-19}$$

At the same point, the reflected current which is propagating *away* from the end, will be

$$i_r\left(t - \frac{L}{2c}\right) = |I| \sin(2\pi f t - \pi L_\lambda) \tag{2-20}$$

At the open end of the conductor the total current, or net charge flow, is zero, because the applied and reflected currents are both positive in the directions in which they propagate, but they propagate in opposite directions.

Thus, the radiated electric field of the *applied* current will be

$$E_{La}(t) = E_L(\theta)\cos\left[2\pi f\left(t - \frac{d}{c}\right) + \pi L_\lambda\right]$$

$$= E_L(\theta)\left\{\cos(\pi L_\lambda)\cos\left[2\pi f\left(t - \frac{d}{c}\right)\right] - \sin(\pi L_\lambda)\sin\left[2\pi f\left(t - \frac{d}{c}\right)\right]\right\}$$

$$\tag{2-21}$$

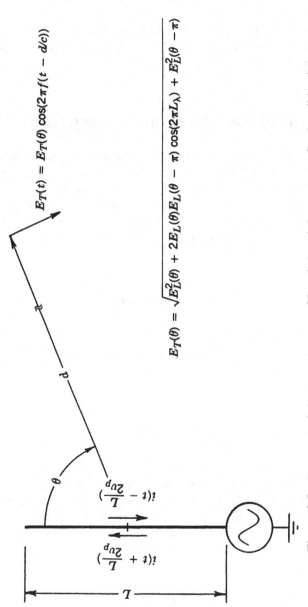

$$E_T(t) = E_T(\theta) \cos(2\pi f(t - d/c))$$

$$E_T(\theta) = \sqrt{E_L^2(\theta) + 2E_L(\theta)E_L(\theta - \pi) \cos(2\pi L_\lambda) + E_L^2(\theta - \pi)}$$

Figure 2-8. An open-ended current segment of length L and the geometry for describing its electric field $E_T(t)$.

And, the radiated electric field of the *reflected* current will be

$$E_{Lr}(t)$$

$$= E_L(\theta - \pi) \cos \left[2\pi f \left(t - \frac{d}{c} \right) - \pi L_\lambda \right]$$

$$= E_L(\theta - \pi) \left\{ \cos(\pi L_\lambda) \cos \left[2\pi f \left(t - \frac{d}{c} \right) \right] \right.$$

$$\left. + \sin(\pi L_\lambda) \sin \left[2\pi f \left(t - \frac{d}{c} \right) \right] \right\} \qquad (2\text{-}22)$$

The total radiated electric field of these two currents will be

$$E_{La}(t) + E_{Lr}(t)$$

$$= [E_L(\theta) + E_L(\theta - \pi)] \cos(\pi L_\lambda) \cos \left[2\pi f \left(t - \frac{d}{c} \right) \right]$$

$$+ [E_L(\theta - \pi) - E_L(\theta)] \sin(\pi L_\lambda) \sin \left[2\pi f \left(t - \frac{d}{c} \right) \right]$$

$$= |E_T(\theta)| \cos \left[2\pi f \left(t - \frac{d}{c} \right) + \psi \right] \qquad (2\text{-}23)$$

where, from A-10, A-11, and A-30 of Appendix A, it follows that

$$|E_T(\theta)| = \sqrt{E_L^2(\theta) + 2E_L(\theta)E_L(\theta - \pi)\cos(2\pi L_\lambda) + E_L^2(\theta - \pi)} \qquad (2\text{-}24)$$

Expressions for $E_T(\theta)$ are rather complicated, but the magnitude of a radiation is the quantity of primary interest here. The magnitude given in Eq. 2-24 is a not-too-complicated general expression for the radiation pattern of $E_{La}(t) + E_{Lr}(t)$. Therefore, the radiation patterns of open-ended current segments of any length can be readily obtained.

For some lengths, $E_T(\theta)$ will be a relatively simple expression. For example, suppose $\cos(2\pi L_\lambda) = \pm 1$. In either case, the quantity under the square root sign in Eq. 2-24 is a perfect square. In other words, if $\cos(2\pi L_\lambda) = +1$, then

$$E_T(\theta) = \sqrt{E_L^2(\theta) + 2E_L(\theta)E_L(\theta - \pi) + E_L^2(\theta - \pi)}$$

$$= \sqrt{[E_L(\theta) + E_L(\theta - \pi)]^2}$$

$$= E_L(\theta) + E_L(\theta - \pi) \tag{2-25}$$

And if $\cos(2\pi L_\lambda) = -1$, then

$$E_T(\theta) = \sqrt{E_L^2(\theta) - 2E_L(\theta)E_L(\theta - \pi) + E_L^2(\theta - \pi)}$$

$$= \sqrt{[E_L(\theta) - E_L(\theta - \pi)]^2}$$

$$= E_L(\theta) - E_L(\theta - \pi) \tag{2-26}$$

Also see Section A.3 of Appendix A.

Example 2-7   Expressions for the radiation patterns of open-ended current segments of lengths equal to 2, 3, 4, 5, 6, 7, and 8 current elements can be found from the expressions for $E_L(\theta)$ and for $E_L(\theta - \pi)$ in Eqs. 2-14 and 2-18 and from the above expressions. For example, if $L_\lambda = 8L_{e\lambda}$ and $L_{e\lambda} = \frac{1}{16}$, then as shown in Section A.3 of Appendix A

$$E_L(t) = [E_L(\theta) - E_L(\theta - \pi)]\sin\left[2\pi f\left(t - \frac{d}{c}\right)\right]$$

$$= \left[\frac{Z_0|I|}{\pi d}\frac{\cos(\pi/2\cos\theta)}{\sin\theta}\right]\sin\left[2\pi f\left(t - \frac{d}{c}\right)\right] \tag{2-27}$$

Cross sections of each of the radiation patterns of open-ended current segments of lengths equal to 2, 3, 4, 5, 6, 7, and 8 current elements of length $L_{e\lambda} = \frac{1}{16}$, together with their component patterns, are shown in Fig. 2-9.

Example 2-8   In this example, the radiation patterns of open-ended current segments equal in length to $4L_e$, $8L_e$, $12L_e$, $16L_e$, $20L_e$, $24L_e$, $28L_3$, and $32L_e$ ($L_\lambda = \frac{1}{4}, \frac{1}{2}, \frac{3}{4}, 1, 1\frac{1}{4}, 1\frac{1}{2}, 1\frac{3}{4}$, and 2 wavelengths) are found. The radiation patterns of their forward currents are among those illustrated in Figs. 2-6 and 2-7. The radiation patterns of the reflected currents are, of course, the same as those of the applied currents after they are rotated 180°. The radiation patterns of the total current in each of these segments, obtained with Eq. 2-24, are shown in Fig. 2-10. These current segments are the equivalents of monopole antennas of the lengths given. The currents in dipole antennas, or adjacent monopoles, and their radiations, are examined in the following section.

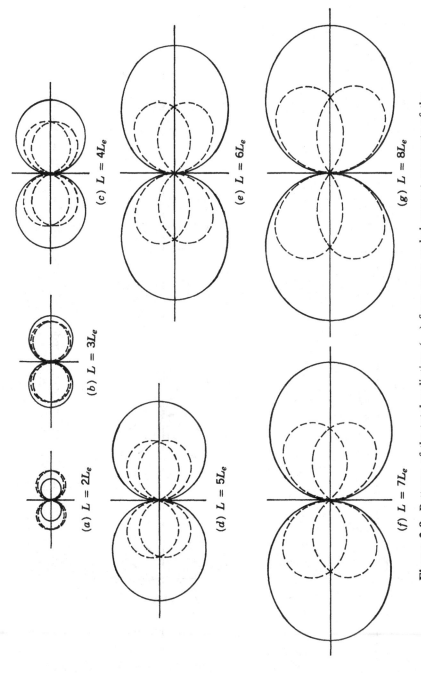

Figure 2-9. Patterns of the total radiation (—) from open-ended current segments of the lengths indicated for the same applied current in each case, and the radiation patterns of the applied and reflected currents (- - -).

(a) $L = 2L_e$

(b) $L = 3L_e$

(c) $L = 4L_e$

(d) $L = 5L_e$

(e) $L = 6L_e$

(f) $L = 7L_e$

(g) $L = 8L_e$

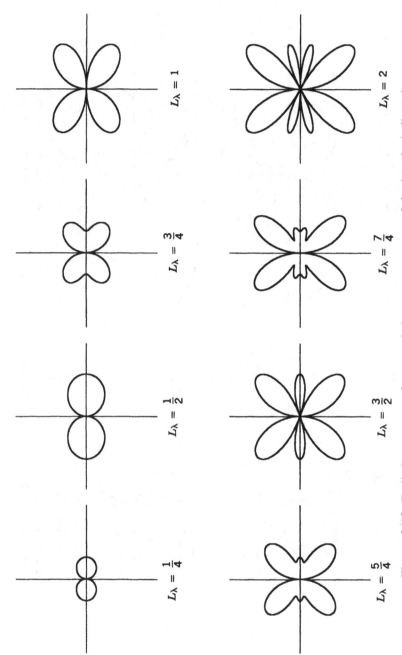

Figure 2-10. Radiation patterns of open-ended current segments of the lengths indicated when the same sinusoidal current is applied in each case.

$L_\lambda = \frac{1}{4}$

$L_\lambda = \frac{1}{2}$

$L_\lambda = \frac{3}{4}$

$L_\lambda = 1$

$L_\lambda = \frac{5}{4}$

$L_\lambda = \frac{3}{2}$

$L_\lambda = \frac{7}{4}$

$L_\lambda = 2$

2.5 RADIATIONS OF DIPOLE ANTENNAS

Consider a pair of equally long, open-ended, parallel conductors that are connected to opposite terminals of a current source, as shown in Fig. 2-11a. At all points in those conductors that are equally distant from the source, the applied currents are equal and oppositely directed. The same is true for the reflected currents. The applied current wave in each conductor propagates from the source to the open end of the conductor, and the reflected current wave in each conductor propagates from the open end of the conductor back to the source. The total currents on each of the two conductors, then, are standing waves of current that are equal in magnitude, but oppositely directed. Because these currents are

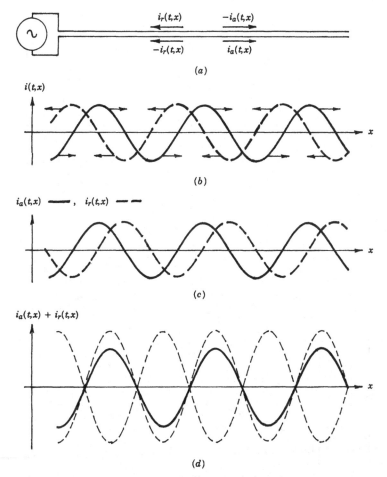

Figure 2-11. Sinusoidal current behavior in an unterminated parallel-wire cable: (a) the cable and its currents; (b) current propagation and reflection; (c) instantaneous applied and reflected currents; (d) instantaneous total current (—) and its maximum values as a standing wave (---).

equal and oppositely directed, so long as their conductors are very close to each other, as shown in Fig. 2-11a, their radiations will cancel.

A dipole antenna consists of two current segments that are open-ended and receive their currents from opposite terminals of the same source. Therefore, a dipole antenna is easily created from two conductors like those shown in Fig. 2-11a. This is done by simply bending each conductor 90° back from its end, as shown in Fig. 2-12a. Each bent-back segment is given length L, so that the current is applied by the conductors at the center of the dipole, and its overall length is $2L$.

The analysis of dipole radiation is based here on that of a current segment when its length is doubled. Referring to Fig. 2-12b, the current that propagates upward at the center of the dipole's upper segment is the applied current

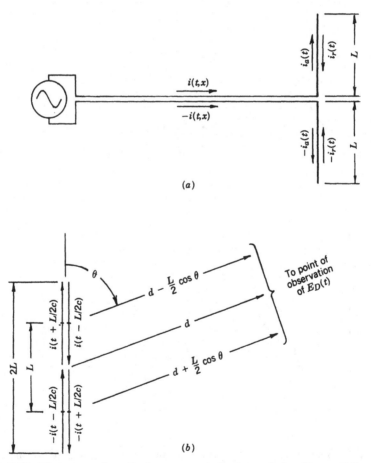

(a)

(b)

Figure 2-12. The currents in a dipole antenna, their phase relationships, and the geometry for finding the radiation pattern: (a) the antenna and its currents; (b) current phases and radiation geometry.

$$i\left(t + \frac{L}{2c}\right) = |I| \sin(2\pi f t + \pi L_\lambda) \tag{2-28}$$

And, the distance to the observation point will be $d - L/2 \cos \theta$. Therefore, the radiated electric field arising from this current will be

$$E_{u1}(t) = E_L(\theta) \cos \left[2\pi f t + \pi L_\lambda - \frac{2\pi f}{c} \left(d - \frac{L}{2} \cos \theta \right) \right]$$

$$= E_L(\theta) \cos \left[2\pi f \left(t - \frac{d}{c} \right) + \pi L_\lambda (\cos \theta + 1) \right] \tag{2-29}$$

The current that propagates upward at the center of the dipole's lower segment is the reflected current

$$-i\left(t - \frac{L}{2c}\right) = -|I| \sin(2\pi f t - \pi L_\lambda) \tag{2-30}$$

The distance to the observation point from the center of this segment, as shown in Fig. 2-11b, will be $d + L/2 \cos \theta$. Therefore, the radiated electric field arising from this current will be

$$E_{u2}(t) = -E_L(\theta) \cos \left[2\pi f t - \pi L_\lambda - \frac{2\pi f}{c} \left(d + \frac{L}{2} \cos \theta \right) \right]$$

$$= -E_L(\theta) \cos \left[2\pi f \left(t - \frac{d}{c} \right) - \pi L_\lambda (\cos \theta + 1) \right] \tag{2-31}$$

From the identity $\cos(a + b) - \cos(a - b) = -2 \sin(a) \sin(b)$, it follows that the total electric field radiated by the upwardly propagating currents is

$$E_{u1}(t) + E_{u2}(t) = E_L(\theta) \cos \left[2\pi f \left(t - \frac{d}{c} \right) + \pi L_\lambda (\cos \theta + 1) \right]$$

$$- E_L(\theta) \cos \left[2\pi f \left(t - \frac{d}{c} \right) - \pi L_\lambda \cos(\theta + 1) \right]$$

$$= -2E_L(\theta) \sin(\pi L_\lambda (\cos \theta + 1)) \sin \left[2\pi f \left(t - \frac{d}{c} \right) \right] \tag{2-32}$$

Similarly, the current that propagates downward in the upper segment is the reflected current

$$i\left(t - \frac{L}{2c}\right) = |I|\sin(2\pi f t - \pi L_\lambda) \tag{2-33}$$

Therefore, its radiated electric field is

$$E_{d1}(t) = E_L(\theta - \pi)\cos\left[2\pi f t - \pi L_\lambda - \frac{2\pi f}{c}\left(d - \frac{L}{2}\cos\theta\right)\right]$$

$$= E_L(\theta - \pi)\cos\left[2\pi f\left(t - \frac{d}{c}\right) + \pi L_\lambda(\cos\theta - 1)\right] \tag{2-34}$$

And, the current that propagates downward in the lower segment is the applied current

$$-i\left(t + \frac{L}{2c}\right) = -|I|\sin(2\pi f t + \pi L_\lambda) \tag{2-35}$$

Therefore, its radiated electric field is

$$E_{d2}(t) = -E_L(\theta - \pi)\cos\left[2\pi f t + \pi L_\lambda - \frac{2\pi f}{c}\left(d + \frac{L}{c}\cos\theta\right)\right]$$

$$= -E_L(\theta - \pi)\cos\left[2\pi f\left(t - \frac{d}{c}\right) - \pi L_\lambda(\cos\theta - 1)\right] \tag{2-36}$$

Using the same identity as before the total electric field radiated by the downwardly propagating currents is seen to be

$$E_{d1}(t) + E_{d2}(t) = E_L(\theta - \pi)\cos\left[2\pi f\left(t - \frac{d}{c}\right) + \pi L_\lambda(\cos\theta - 1)\right]$$

$$- E_L(\theta - \pi)\cos\left[2\pi f\left(t - \frac{d}{c}\right) - \pi L_\lambda\cos(\theta - 1)\right]$$

$$= -2E_L(\theta - \pi)\sin\left[\pi L_\lambda(\cos\theta - 1)\right]\sin\left[2\pi f\left(t - \frac{d}{c}\right)\right] \tag{2-37}$$

It follows, then, by adding Eqs. 2-32 and 2-37, that a dipole of length $2L$ has a radiated electric field

$$E_D(t) = E_{u1}(t) + E_{u2}(t) + E_{d1}(t) + E_{d2}(t)$$

$$= E_D(\theta) \sin\left[2\pi f\left(t - \frac{d}{c}\right)\right]$$

where

$$E_D(\theta) = -2\{E_L(\theta) \sin[\pi L_\lambda(\cos\theta + 1)] + E_L(\theta - \pi) \sin[\pi L_\lambda(\cos\theta - 1)]\} \quad (2\text{-}38)$$

Thus, with $E_L(\theta)$ and $E_L(\theta - \pi)$ as given in Eqs. 2-14 and 2-18, and $v_p = c$,

$$E_D(\theta) = \frac{Z_0|I|}{\pi d} \frac{\sin\theta}{\cos\theta + 1} \sin[\pi L_\lambda(\cos\theta + 1)] \sin[\pi L_\lambda(\cos\theta - 1)]$$

$$- \frac{Z_0|I|}{\pi d} \frac{\sin\theta}{\cos\theta - 1} \sin[\pi L_\lambda(\cos\theta + 1)] \sin[\pi L_\lambda(\cos\theta - 1)]$$

$$= \frac{Z_0|I|}{\pi d} \left(\frac{2\sin\theta}{\cos^2\theta - 1}\right)$$

$$\cdot [\sin^2(\pi L_\lambda \cos\theta) \cos^2(\pi L_\lambda) - \cos^2(\pi L_\lambda \cos\theta) \sin^2(\pi L_\lambda)]$$

$$= \frac{Z_0|I|}{\pi d} \frac{[1 - \cos(2\pi L_\lambda \cos\theta)] \cos^2(\pi L_\lambda)}{\sin\theta}$$

$$- \frac{Z_0|I|}{\pi d} \frac{[1 + \cos(2\pi L_\lambda \cos\theta)] \cos^2(\pi L_\lambda)}{\sin\theta}$$

$$= \frac{Z_0|I|}{\pi d} \frac{\cos^2(\pi L_\lambda) - \sin^2(\pi L_\lambda) - \cos(2\pi L_\lambda \cos\theta)}{\sin\theta}$$

$$= \frac{Z_0|I|}{\pi d} \frac{\cos(2\pi L_\lambda) - \cos(2\pi L_\lambda \cos\theta)}{\sin\theta} \quad (2\text{-}39)$$

all of which follows from trigonometric identities given in Appendix A. Equation 2-39 is the pattern factor of a dipole of any length, where $2L_\lambda$ is its overall length.

Example 2-9 The eight dipole radiation patterns shown in Fig. 2-13 each have pattern factors that were obtained from Eq. 2-39. For example, when the overall length of the dipole is $2L_\lambda = \frac{1}{2}$ wavelength, then

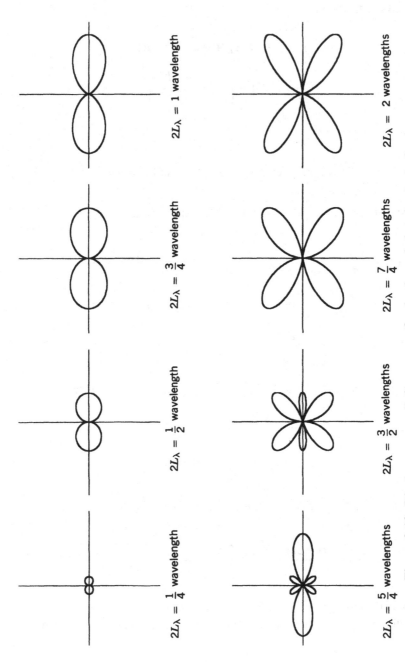

Figure 2-13. Dipole antenna radiation patterns for the tip-to-tip lengths shown for the same applied sinusoidal current in each case.

$2L_\lambda = \frac{1}{4}$ wavelength

$2L_\lambda = \frac{1}{2}$ wavelength

$2L_\lambda = \frac{3}{4}$ wavelength

$2L_\lambda = 1$ wavelength

$2L_\lambda = \frac{5}{4}$ wavelengths

$2L_\lambda = \frac{3}{2}$ wavelengths

$2L_\lambda = \frac{7}{4}$ wavelengths

$2L_\lambda = 2$ wavelengths

$$E_D(\theta) = \frac{Z_0|I|}{\pi d} \frac{\cos(2\pi L_\lambda) - \cos[2\pi L_\lambda \cos\theta)]}{\sin\theta}$$

$$= \frac{Z_0|I|}{\pi d} \frac{\cos(\pi/2) - \cos[(\pi/2)\cos\theta]}{\sin\theta}$$

$$= -\frac{Z_0|I|}{\pi d} \frac{\cos[(\pi/2)\cos\theta]}{\sin\theta} \tag{2-40}$$

The pattern factors of each of the radiation patterns illustrated in Fig. 2-13 were obtained in this way, and the same applied current was assumed in each case.

2.6 RADIATIONS OF FOLDED DIPOLE ANTENNAS

The radiations of some dipoles can be increased by also folding the segments that are bent back at the ends of the conductors that deliver the applied current. One way to fold the segments is that shown in Fig. 2-14. Notice that, because of the folds, there will be two sets of current that propagate upward, and two sets that propagate downward. There are, in effect, then, two dipoles at the same location carrying currents that differ only in phase. The applied currents radiate as one dipole, the reflected currents radiate as another, and the folded dipole's radiation is the sum of the two.

As seen from Fig. 2-14, the reflected currents lead the applied currents in phase by $2L_\lambda$. Therefore, when the radiated electric field of the applied currents is

$$E_a(t) = E_D(\theta)\sin\left[2\pi f\left(t - \frac{d}{c}\right) + 2\pi L_\lambda\right] \tag{2-41}$$

the radiated electric field of the reflected currents is

$$E_r(t) = E_D(\theta)\sin\left[2\pi f\left(t - \frac{d}{c}\right) - 2\pi L_\lambda\right] \tag{2-42}$$

And the total radiated electric field from the folded dipole will be

$$E_F(t) = E_a(t) + E_r(t)$$

$$= E_F(\theta)\sin\left[2\pi f\left(t - \frac{d}{c}\right)\right]$$

where

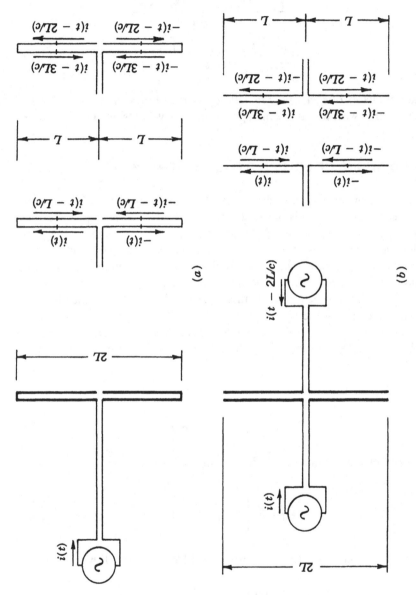

Figure 2-14. The equivalency of (*a*) a folded dipole and (*b*) a pair of dipoles of the same length with appropriately phased currents.

38

$$E_F(\theta) = 2E_D(\theta)\cos(2\pi L_\lambda) \qquad (2\text{-}43)$$

This follows from trigonometric identity A-5 that says, $\sin(a+b) + \sin(a-b) = 2\sin(a)\cos(b)$, for any a and b.

Thus, whenever $\cos(2\pi L_\lambda) = \pm 1$, then $|E_F(\theta)| = 2|E_D(\theta)|$ and the folded dipole has twice the radiated electric field of an unfolded dipole of the same length. On the other hand, if $L_\lambda = \frac{1}{4}$, for example, then $\cos(2\pi L_\lambda) = \cos(\pi/2) = 0$, and there is effectively no radiation from the folded dipole. This is another illustration of the importance of knowing both a current's path length and its path geometry in determining how effectively it will radiate.

Finally, consider the folded dipole illustrated in Fig. 2-15. Here there are, in effect, three dipoles at the same location. The currents in the second dipole lead those of the first dipole in phase by $2L_\lambda$, and the currents in the third dipole lead those of the first dipole in phase by $4L_\lambda$. The currents in the second dipole differ in sign from the currents of the other two. Thus, the radiated electric fields of the three sets of dipole currents are

$$E_{D1}(t) = E_D(\theta)\sin\left[2\pi f\left(t - \frac{d}{c}\right) + 4\pi L_\lambda\right]$$

$$E_{D2}(t) = -E_D(\theta)\sin\left[2\pi f\left(t - \frac{d}{c}\right)\right]$$

and

$$E_{D3}(t) = E_D(\theta)\sin\left[2\pi f\left(t - \frac{d}{c}\right) - 4\pi L_\lambda\right] \qquad (2\text{-}44)$$

From trigonometric identity A-5, used above, it follows that

$$E_{D1}(t) + E_{D3}(t) = 2E_D(\theta)\cos(4\pi L_\lambda)\sin\left[2\pi f\left(t - \frac{d}{c}\right)\right] \qquad (2\text{-}45)$$

Thus, the radiated electric field of this folded dipole is

$$E_{FD}(t) = [E_{D1}(t) + E_{D3}(t)] + E_{D2}(t)$$

$$= E_{FD}(\theta)\sin\left[2\pi f\left(t - \frac{d}{c}\right)\right]$$

where

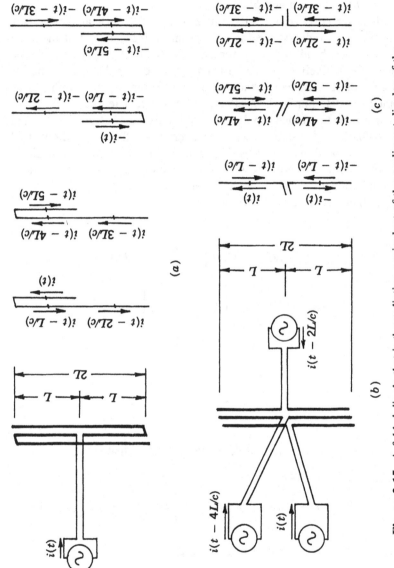

Figure 2-15. A folded dipole that is the radiating equivalent of three adjacent dipoles of the same length when their currents are appropriately phased: (*a*) a folded dipole and its applied and reflected currents; (*b*) a three-dipole equivalent; (*c*) the dipole currents.

$$E_{FD}(\theta) = E_D(\theta)[2\cos(4\pi L_\lambda) - 1] \tag{2-46}$$

These folds will increase the radiation of the dipole most whenever its length L_λ is such that $\cos(4\pi L_\lambda) = -1$, because then $|E_{FD}(\theta)| = 3|E_D(\theta)|$. For example, if L_λ is any odd multiple of $\frac{1}{4}$, then $\cos(4\pi L_\lambda) = -1$ and the radiated electric field of an unfolded dipole of the same length will be tripled. However, if L_λ is an even multiple of $\frac{1}{4}$, then $\cos(4\pi L_\lambda) = +1$, and the folding has no effect, because $|E_{FD}(\theta)| = |E_D(\theta)|$.

2.7 SUMMARY

This chapter is meant to provide an understanding of the basic causes of electromagnetic radiation and how to control it. Among the concepts discussed are antenna currents, current segments that behave like antenna currents, and the radiation patterns of both. An *antenna current* is a current, or portion of a current, that is viewed as a radiator, whether it is supposed to radiate or not. A *current segment* is a straight-line portion of a current that is isolated conceptually to study its radiation. A *radiation pattern* is a graphical description of the relative magnitude of a current's radiation in any given direction.

The magnitude of the radiation of a segment of sinusoidal current is directly proportional to the frequency and amplitude of that current. The size and shape of the radiation pattern of a sinusoidal current segment are both functions of its length relative to the current's wavelength. Longer current segments generally have larger, more complex radiation patterns. Cross sections of the radiation patterns of segments of current of equal amplitude and frequency, but different lengths, are illustrated in Fig. 2-16. Note that in some directions the increase in the magnitude of the radiation is proportional to the increase in segment length, while in other directions it is not. Nevertheless, it is clear from these examples that one way to reduce the radiation of some current segments is to shorten them.

The size and shape of a current's radiation pattern can also be affected by sudden changes in the impedance the current sees along its path as illustrated in Fig. 2-17. A conductor with a sinusoidal current source at one end and nothing at its other end has two currents in it—the one supplied by the source and a reflection of the source current which occurs at the open end of the conductor. The applied and reflected currents will always be equal and oppositely directed at the point of reflection. Thus, at the open end of the conductor, where there is nothing connected, the current will always be zero. However, at the vast majority of points elsewhere along the conductor, the two currents will not always be equal and they will not always be oppositely directed. Therefore, the total current at those points will seldom be zero. In fact, the total current along such a conductor will be a standing wave of current with twice the amplitude of the

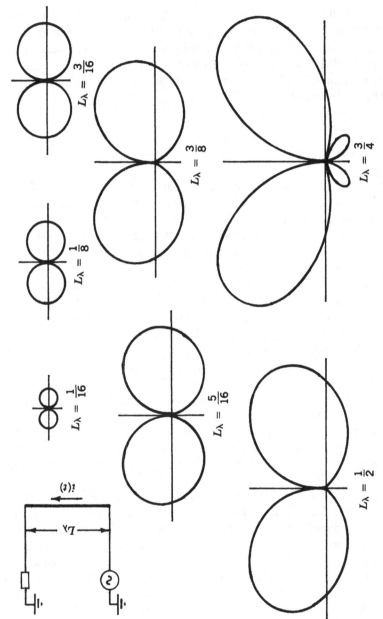

Figure 2-16. Radiation patterns of current segments of the lengths indicated (in fractions of a wavelength) for the same sinusoidal current $i(t)$ in each case.

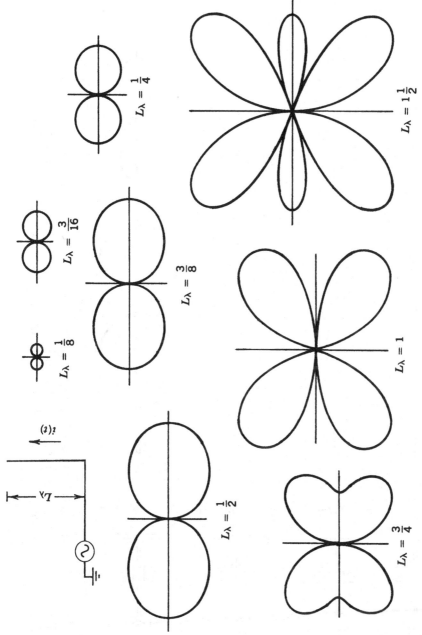

Figure 2-17. Radiation patterns of open-ended current segments of the lengths indicated (in fractions of a wavelength) for the same sinusoidal current, $i(t)$, in each case.

43

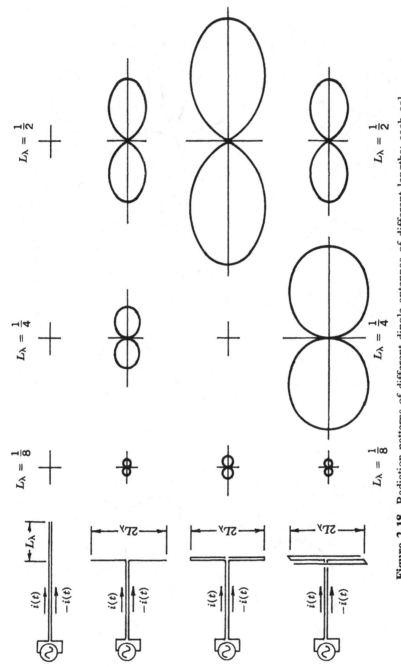

Figure 2-18. Radiation patterns of different dipole antennas of different lengths: each column of patterns is for different antennas of the same length, each row of patterns is for the same kind of antenna of different lengths. The sinusoidal current applied is the same in every case.

applied current. In other words, the radiation of an open-ended current segment will be the radiation of the sum of the applied and reflected currents.

Differences in the radiation patterns of different dipole antennas show that the geometry of a current's path is extremely important in controlling its radiation. A closely parallel pair of open-ended conductors of equal length, connected to opposite terminals of a current source, will radiate little or nothing. However, when those same parallel conductors are folded back at the ends to form a dipole, they can be made to radiate very effectively. From this, it is clear that the path of a time-varying current largely determines the extent with which it radiates.

The radiation patterns of several different types of dipole antennas of different lengths, each supplied with the same current, are illustrated in Fig. 2-18. The extreme differences in the sizes of those radiation patterns are due only to differences in the paths of the currents that cause them, which clearly illustrates the importance of current path geometry in controlling radiation.

The ideas presented in this chapter are all developed in greater detail for circuit currents in the chapters that follow.

CHAPTER 3

SINUSOIDAL CIRCUIT-CURRENT RADIATIONS

3.1 INTRODUCTION

For those readers whose primary interest is the effective reduction of unnecessary electromagnetic radiation, this will be a key chapter. Circuit-current radiations are characterized here with simple mathematics, and the primary concepts on which to base the reduction and control of circuit-current radiations are identified. Also, several examples of the radiations of sinusoidal circuit currents are examined graphically.

To minimize the radiation of a rectangular circuit current, the path width W of the current's circuital path must be minimized. The path width of a circuit is illustrated in Fig. 3-1. It is the distance between the connecting-current paths from source to load. The maximum radiations of all of the rectangular circuit currents examined in this chapter are directly proportional to their path widths. The radiations of circuit currents depend on several other parameters as well, but no other parameter is as freely adjustable as the path of a circuit current. That is an additional reason path width is important in effectively controlling circuit-current radiations.

Nonrectangular circuits and other parameters that affect the radiations of periodic circuit currents, which are sums of sinusoidal components, are examined in Chapter 6. It is again emphasized there, however, that properly reducing the path width of any circuit current is the most significant step that can be taken to reduce and control its radiations.

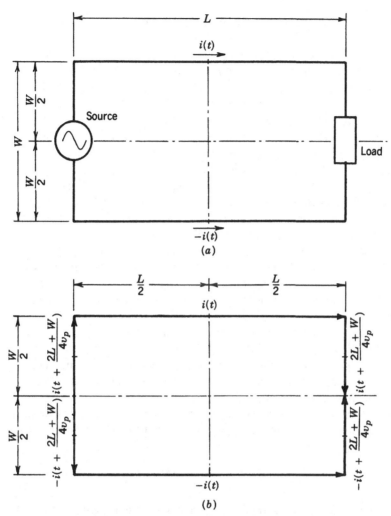

Figure 3-1. (*a*) A rectangular circuit and (*b*) the current-segment model of its current.

3.2 CIRCUIT CURRENTS

Consider the rectangular circuit illustrated in Fig. 3-1. Each straight-line section of that circuit is a current segment, such as those discussed in the previous chapter. In these current segments there are assumed to be no reflected currents. However, phase differences do exist in the currents of the different segments of a circuit. Phase differences in currents will be evaluated, as before, based on the assumption that currents propagate from source to load with a speed v_p, which approaches the speed of light, $c = 3 \times 10^8$ meters/second. It is important to include these phase differences in predicting the radiation patterns of circuit currents, no matter how small the circuit may be. Phase differences in circuit currents have substantial effects on the radiations of small circuits, as well as large circuits.

As in Fig. 3-1a, the direction of a positive or negative current will be indicated by an arrow beside the conductor and the notation for the current at that point. Otherwise, as in Fig. 3-1b, a current's direction will be the same as the direction in which its phase propagates, as indicated by the arrowhead on its segment. The phase given for a current segment will be that at the center of the segment, and that will be the phase assumed in calculating the radiated electric field. And, of course, the sign of a current tells whether positive or negative charge flows in the direction indicated.

As shown in Fig. 3-1b, positive and negative currents propagate from the center of the source to a point equally distant from the source along their two paths. Those paths are often referred to as forward (positive) current and return (negative) current paths; here they will usually be referred to as connecting-current paths. Note that unless a given circuit has a perimeter longer than one wavelength, the connecting currents will be the same in phase and direction only at the center of the source and at the halfway point—the point that is equally distant from the center of the source along either of the two connecting-current paths. Also note that although the load will often be shown to be centered on the half-way point, it need not be. That point is completely independent of the position of the load. The load determines the same relative amplitudes and phases of the voltage and current wherever it is positioned in the circuit. The position of the load relative to the source has no effect whatsoever on differences in the current's phase at different points in the circuit.

As shown in Fig. 3-1, at the center of the upper horizontal current segment of length L, the current that propagates from left to right is assumed to be

$$i(t) = |I| \sin(2\pi ft) \tag{3-1}$$

Therefore, at the center of the lower horizontal current segment of length L, an equal distance from the source, the current that propagates from left to right is

$$-i(t) = -|I| \sin(2\pi ft) \tag{3-2}$$

the negative of the current in the upper segment.

The time required for current to propagate from the center of the source to the center of either of those horizontal segments is $t_p = (W/2 + L/2)/v_p$ seconds, since v_p is the propagation velocity of the current in the circuit. Thus, the wavelength of a periodic current of frequency f will be v_p/f, the wavelength of its radiation will be c/f, and $v_p/f \leq c/f$.

Therefore, given the currents of Eqs. 3-1 and 3-2 and accounting for phase differences, the current propagating upward at the center of the source in the circuit of Fig. 3-1 will be

$$i(t + t_p) = |I| \sin\left[2\pi f \left(t + \frac{L + W}{2v_p} \right) \right]$$

$$= |I| \sin[2\pi ft + \pi(L'_\lambda + W'_\lambda)] \qquad (3\text{-}3)$$

And the current propagating downward at the source center will be $-i(t + t_p)$. In the latter of Eqs. 3-3, $L'_\lambda = Lf/v_p$, and $W'_\lambda = Wf/v_p$, are the current segment lengths in fractions of the current's wavelength. Since W, L, and v_p/f, all have the same units, L'_λ and W'_λ have no units.

Similarly, the time required for the current to propagate from the center of either horizontal segment to the center of the right-hand vertical segment, the half-way point, is also $t_p = (W/2) + L/2)/v_p$ seconds. Therefore, the current propagating downward at the center of the right-hand segment of length W is

$$i(t - t_p) = |I| \sin\left[2\pi f \left(t - \frac{L + W}{2v_p} \right) \right]$$

$$= |I| \sin[2\pi ft - \pi(L'_\lambda + W'_\lambda)] \qquad (3\text{-}4)$$

And the current propagating upward at the segment center will be $-i(t - t_p)$, because the current is the same at that point.

The radiations of these current segments will now be characterized. Based on those results, a general description of the radiation pattern of a rectangular circuit current will be developed. Finally, the radiation patterns of several sinusoidal circuit currents in each of the three coordinate planes will be determined, first for small rectangular circuits, then for larger rectangular circuits. Circuits of other shapes are discussed in Chapter 6.

3.3 RADIATION AND CURRENT SEGMENT ORIENTATION

Recall that any segment of current $i(t) = |I| \sin(2\pi ft)$ of length L causes a radiated electric field $E_L(t) = E_L(\theta) \cos[2\pi f (t - d/c)]$ a distance d from its center. In this expression, $E_L(\theta)$ is the *radiation pattern factor*, the magnitude of which yields the radiation pattern of $E_L(t)$; θ is the *observation angle*, measured from the axis of the current segment to the line from the center of the segment to the observation point of $E_L(t)$; and d is the *observation distance*, the length of the line from the center of the current segment to the observation point of $E_L(t)$. The distance d will typically be 10 meters, or more, so that, for most frequencies of interest, $d \gg c/f$. As shown in Fig. 3-2, if the positive direction of a segment of current $i(t)$ is upward, then the positive direction of its electric field $E_L(t)$ is defined to be clockwise to the right of the segment in the plane that contains it, and counterclockwise to the left. In the plane through the seg-

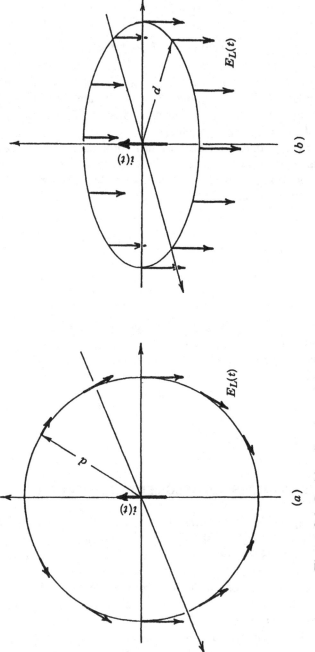

Figure 3-2. Positive directions of the radiated electric field $E_L(t)$ of a current segment of length L relative to the positive direction of its current $i(t)$: (a) current segment in the plane of observation of $E(t)$; (b) current segment normal to the plane of observation of $E(t)$.

ment's center, and to which it is normal, the positive direction of $E_L(t)$ will be downward.

The current segments of circuit currents are not all collinear as those of the antennas discussed in the previous chapter. However, in any given plane their radiation pattern factors are all functions of the same angle. Therefore, expressions describing those pattern factors will differ depending on the orientation of a given current segment, the direction of its positive current, the direction in which the current propagates, and the angle of observation.

For example, a current segment lying on the x axis, as shown in Fig. 3-3a, with a positive current and a phase that propagates with velocity v_p, both in the positive x direction, has the xy-plane radiation pattern factor

$$E_L(\phi) = \frac{Z_0|I|}{2\pi d} \sin \phi \; \frac{\sin[\pi L_\lambda(\cos \phi - c/v_p)]}{\cos \phi - c/v_p} \tag{3-5a}$$

And if $L_\lambda \to 0$, so that $0 < L_\lambda = L_{e\lambda} \leq \frac{1}{16}$, and $v_p \to c$, so that $\sin[\pi L_\lambda(\cos \phi - c/v_p)] \cong \pi L_\lambda(\cos \phi - c/v_p)$, the current segment is a current element and

$$E_L(\phi) = E_e(\phi) = \frac{Z_0|I|L_{e\lambda}}{2d} \sin \phi \tag{3-5b}$$

A current segment lying on the y axis, as illustrated in Fig. 3-3b, with a positive current that is directed and propagates with velocity v_p in the positive y direction has the xy-plane radiation pattern factor

$$E_L\left(\phi - \frac{\pi}{2}\right) = \frac{Z_0|I|}{2\pi d} \sin\left(\phi - \frac{\pi}{2}\right) \frac{\sin\{\pi L_\lambda[\cos(\phi - \pi/2) - c/v_p]\}}{\cos(\phi - \pi/2) - c/v_p}$$

$$= \frac{-Z_0|I|}{2\pi d} \cos \phi \; \frac{\sin[\pi L_\lambda(\sin \phi - c/v_p)]}{\sin \phi - c/v_p} \tag{3-6a}$$

If the length of the segment is $0 < L_\lambda = L_{e\lambda} \leq \frac{1}{16}$, then it is a current element; and if $v_p \to c$, then

$$E_L\left(\phi - \frac{\pi}{2}\right) = E_e\left(\phi - \frac{\pi}{2}\right) = \frac{-Z_0|I|L_{e\lambda}}{2d} \cos \phi \tag{3-6b}$$

A current segment lying on the x axis, as illustrated in Fig. 3-3c, with a positive current that is directed and propagates with velocity v_p in the negative x direction has the xy-plane radiation pattern factor

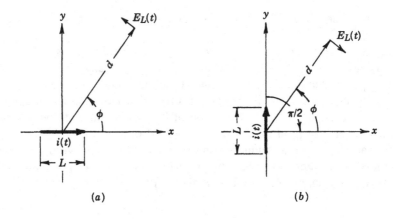

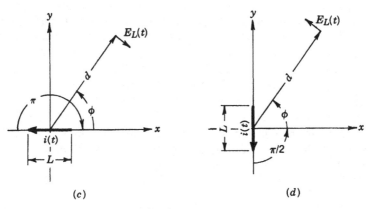

Figure 3-3. Different current-segment orientations and the angles with which their radiation pattern factors are defined in the xy plane.

$$E_L(\phi - \pi) = \frac{Z_0|I|}{2\pi d} \sin(\phi - \pi) \frac{\sin\{\pi L_\lambda [\cos(\phi - \pi) - c/v_p]\}}{\cos(\phi - \pi) - c/v_p}$$

$$= \frac{-Z_0|I|}{2\pi d} \sin\phi \frac{\sin[\pi L_\lambda(\cos\phi + c/v_p)]}{\cos\phi + c/v_p} \qquad (3\text{-}7a)$$

And if $v_p \rightarrow c$ and $0 < L_\lambda = L_{e\lambda} \leq \frac{1}{16}$, then the segment is a current element and

$$E_L(\phi - \pi) = E_e(\phi - \pi) = \frac{-Z_0|I|L_{e\lambda}}{2d} \sin\phi \qquad (3\text{-}7b)$$

A current segment lying on the y axis, as illustrated in Fig. 3-3d, with a positive current that is directed and propagates with velocity v_p in the negative y direction has the xy-plane radiation pattern factor

$$E_L\left(\phi + \frac{\pi}{2}\right) = \frac{Z_0|I|}{2\pi d}\sin\left(\phi + \frac{\pi}{2}\right)\frac{\sin\{\pi L_\lambda[\cos(\phi + \pi/2) - c/v_p]\}}{\cos(\phi + \pi/2) - c/v_p}$$

$$= \frac{Z_0|I|}{2\pi d}\cos\phi\,\frac{\sin[\pi L_\lambda(\sin\phi + c/v_p)]}{\sin\phi + c/v_p} \tag{3-8a}$$

And if $v_p \rightarrow c$ and the length of the segment is $0 < L_\lambda = L_{e\lambda} \leq \frac{1}{16}$, then it is a current element and

$$E_L\left(\phi + \frac{\pi}{2}\right) = E_e\left(\phi + \frac{\pi}{2}\right) = \frac{Z_0|I|L_{e\lambda}}{2d}\cos\phi \tag{3-8b}$$

Thus are the radiation pattern factors of differently oriented current segments and current elements described as functions of the same angle.

3.4 CIRCUIT-CURRENT RADIATION PATTERNS

The radiation patterns of circuit currents are considerably more complex in shape than those of the antenna currents discussed in the previous chapter. The patterns considered there were symmetrical about one axis, because, for all practical purposes, their current segments were all on the same axis. As a result, each radiation pattern was symmetrical about that axis, and the pattern was the same in any plane that contained that axis. However, circuit-current segments are not all parallel to the same axis, because they form a circuit, or loop. Therefore, the radiation patterns of rectangular circuit currents are obtained here by assuming the circuits to be centered on, and lying in, the xy plane. Their connecting segments are parallel to the x axis, and the segments containing the source and load are parallel to the y axis. Their radiation patterns are described by finding their intersections with the xy plane, the xz plane, and the yz plane.

This approach simplifies the description of the radiation patterns of circuit currents, because the field components found will either lie in, or normal to, one of the coordinate planes. Therefore, at observation points in those planes, the vector components of the total radiated electric field, will either have the same, or opposite, directions, or they will be at right angles to each other. Thus, their magnitudes can easily be found without using vector analysis.

3.4.1   The *xy*-Plane Radiation Pattern

In general, as shown in Fig. 3-1*b*, it takes six current segments to model a rectangular circuit current. However, if the circuit width W, the frequency f, and the phase velocity v_p are such that $0 < W'_\lambda = W f/v_p \leq \frac{1}{16}$, then the current segment pairs at either end of the circuit can each be replaced by a single current element. This can be done, because, as seen from Eqs. 3-5b, 3-6b, 3-7b, and 3-8b, elements of current with opposite signs, propagating in opposite directions, will have the same radiation pattern factor. For example, $-E_e(\phi - \pi/2) = E_e(\phi + \pi/2)$. Thus, the radiations of equal current elements that are positive in opposite directions and propagate in opposite directions will be the same, and their radiations will add. If the sum of their lengths is less than or equal to that of a current element, their total radiation will be that of a single current element with a length equal to that sum.

Notice in Fig. 3-4*a* that as $W \to 0$ in a rectangular circuit, the connecting current segments get closer and closer together. Since they have currents of equal magnitude and opposite sign that propagate in the same direction, as W approaches zero and they approach each other, their total radiation approaches zero. Notice in Fig. 3-4*b* that if W is large and $L \to 0$, then the circuit current pattern approaches that of the applied and reflected currents of a dipole antenna. In other words, if L is made small but W is large, then the circuit current will probably be an effective radiator.

Thus, L can be large, but W should be small to minimize circuit-current radiations. Because of that, and because it simplifies the discussion, it will be assumed for all circuits examined below that $0 < W'_\lambda = W f/v_p \leq \frac{1}{16}$.

From Fig. 3-5, then, it is seen that, in the xy plane, the radiated electric field of the vertical current element on the source side of the circuit of Fig. 3-1 is

$$E_s(t) = E_W\left(\phi - \frac{\pi}{2}\right)\cos\left\{2\pi f\left[t + \frac{L+W}{2v_p} - \frac{1}{c}\left(d + \frac{L}{2}\cos\phi\right)\right]\right\}$$

$$= E_W\left(\phi - \frac{\pi}{2}\right)\cos\left[2\pi f\left(t - \frac{d}{c}\right) - \pi L_\lambda \cos\phi + \pi(L'_\lambda + W'_\lambda)\right]$$

$$\tag{3-9}$$

This expression results from the general expression $E_L(t) = E_L(\theta)\cos[2\pi f(t - d/c)]$. However, the observation distance is now $d + (L/2)\cos\phi$, the current at the center of the segment is now $i(t + t_p) = |I|\sin\{2\pi f[t + (L + W)/2v_p]\}$, and ϕ is measured from the x axis so that the pattern factor is $E_W(\phi - \pi/2)$. Also, $W_\lambda = W f/c$, and, as previously defined, $L'_\lambda = L f/v_p$ and $W'_\lambda = W f/v_p$.

Also from Fig. 3-5, it is seen that, in the xy plane, the radiated electric field of the right-hand current element of the circuit of Fig. 3-1 is

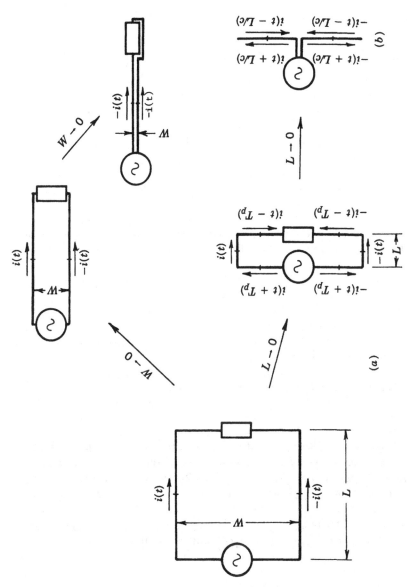

Figure 3-4. As W is made smaller, connecting-current radiations cancel each other more and more; as L is made smaller, the circuit current path becomes more and more like that of a dipole antenna: (*a*) several circuits of length L and width W; (*b*) dipole antenna.

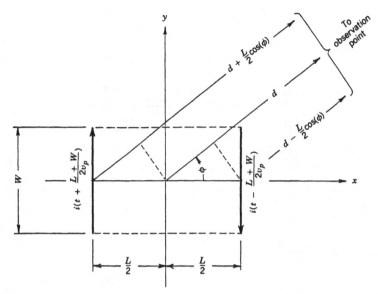

Figure 3-5. The geometry for finding the contribution of the source and load current segments to the xy-plane pattern factor of the radiated electric field of the circuit of Fig. 3-1.

$$E_r(t) = E_W\left(\phi - \frac{\pi}{2}\right) \cos\left\{2\pi f\left[t - \frac{L+W}{2v_p} - \frac{1}{c}\left(d - \frac{L}{2}\cos\phi\right)\right]\right\}$$

$$= -E_W\left(\phi - \frac{\pi}{2}\right) \cos\left[2\pi f\left(t - \frac{d}{c}\right) + \pi L_\lambda \cos\phi - \pi(L'_\lambda + W'_\lambda)\right] \quad (3\text{-}10)$$

This expression also follows from the general expression $E_L(t)$ given above. However, the observation distance is now $d - (L/2)\cos\phi$, the current at the center of the element is now $i(t - t_p) = |I|\sin\{2\pi f[t - (L + W)/2v_p]\}$, and the pattern factor is $E_W(\phi + \pi/2)$. Also, from Eqs. 3-6b and 3-8b, $E_W(\phi + \pi/2) = -E_W(\phi - \pi/2)$.

Therefore, from the identity $\cos(a - b) - \cos(a + b) = 2\sin(a)\sin(b)$, with $a = 2\pi f(t - d/c)$ and $b = \pi L_\lambda \cos\phi - \pi(L'_\lambda + W'_\lambda)$, it follows that

$$E_s(t) + E_r(t) = E_W\left(\phi - \frac{\pi}{2}\right) \cos\left[2\pi f\left(t - \frac{d}{c}\right) - \pi L_\lambda \cos\phi + \pi(L'_\lambda + W'_\lambda)\right]$$

$$- E_W\left(\phi - \frac{\pi}{2}\right) \cos\left[2\pi f\left(t - \frac{d}{c}\right) + \pi L_\lambda \cos\phi - \pi(L'_\lambda + W'_\lambda)\right]$$

$$= 2E_W \left(\phi - \frac{\pi}{2} \right) \sin \left[\pi L_\lambda \cos \phi - \pi(L'_\lambda + W'_\lambda) \right] \sin \left[2\pi f \left(t - \frac{d}{c} \right) \right]$$

$$= \frac{Z_0 |I| W_\lambda}{d} \cos \phi \sin[\pi(L'_\lambda + W'_\lambda - L_\lambda \cos \phi)] \sin \left[2\pi f \left(t - \frac{d}{c} \right) \right]$$

$$= E_V(\phi) \sin \left[2\pi f \left(t - \frac{d}{c} \right) \right] \tag{3-11}$$

Next, referring to Fig. 3-6, it is seen that the observation distance from the center of the upper current segment of length L in the circuit of Fig. 3-1 is $d - (W/2) \sin \phi$. Also, the current at the center of that segment is $i(t) = |I| \sin(2\pi f t)$. Therefore, the radiated electric field from the upper horizontal current segment of the circuit will be

$$E_u(t) = E_L(\phi) \cos \left\{ 2\pi f \left[t - \frac{1}{c} \left(d - \frac{W}{2} \sin \phi \right) \right] \right\}$$

$$= E_L(\phi) \cos \left[2\pi f \left(t - \frac{d}{c} \right) + \pi W_\lambda \sin \phi \right] \tag{3-12}$$

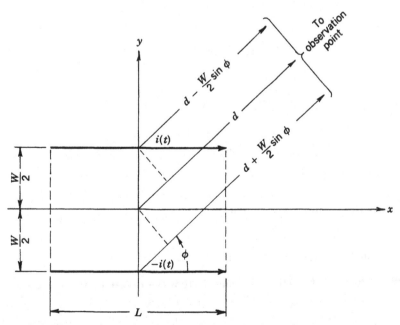

Figure 3-6. The geometry for finding the contribution of the connecting-current segments to the xy-plane pattern factor of the radiated electric field of the circuit of Fig. 3-1.

Similarly, at the distance $d + (W/2) \sin \phi$, the radiated electric field from the lower horizontal segment, with the current $-i(t) = -|I| \sin(2\pi f t)$ at its center, will be

$$E_\ell(t) = -E_L(\phi) \cos\left\{2\pi f\left[t - \frac{1}{c}\left(d + \frac{W}{2}\sin\phi\right)\right]\right\}$$

$$= -E_L(\phi) \cos\left[2\pi f\left(t - \frac{d}{c}\right) - \pi W_\lambda \sin\phi\right] \qquad (3\text{-}13)$$

$E_u(t)$ and $E_\ell(t)$ have the same pattern factor with opposite signs, because the currents that cause them have opposite signs and propagate in the same direction.

Using the trigonometric identity $\cos(a + b) - \cos(a - b) = -2 \sin(a) \sin(b)$, with $a = 2\pi f(t - d/c)$ and $b = \pi W_\lambda \sin \phi$, it can be seen that

$$E_H(t) = E_u(t) + E_\ell(t)$$

$$= E_L(\phi) \cos\left[2\pi f\left(t - \frac{d}{c}\right) + \pi W_\lambda \sin\phi\right]$$

$$\quad - E_L(\phi) \cos\left[2\pi f\left(t - \frac{d}{c}\right) - \pi W_\lambda \sin\phi\right]$$

$$= -2E_L(\phi) \sin(\pi W_\lambda \sin\phi) \sin\left[2\pi f\left(t - \frac{d}{c}\right)\right]$$

$$= E_H(\phi) \sin\left[2\pi f\left(t - \frac{d}{c}\right)\right] \qquad (3\text{-}14)$$

And from Eq. 3-5a it is seen that

$$E_H(\phi) = -2E_L(\phi) \sin(\pi W_\lambda \sin\phi)$$

$$= -2\left\{\frac{Z_0|I|}{2\pi d}\sin\phi\ \frac{\sin[\pi L_\lambda(\cos\phi - c/v_p)]}{\cos\phi - c/v_p}\right\}\sin(\pi W_\lambda \sin\phi)$$

$$= \frac{-Z_0|I|W_\lambda}{d}\ \frac{\sin^2\phi\ \sin[\pi L_\lambda(\cos\phi - c/v_p)]}{\cos\phi - c/v_p} \qquad (3\text{-}15)$$

The latter expression follows because $W_\lambda \le W_\lambda' \le \frac{1}{16}$, which, for all practical purposes, makes $\sin(\pi W_\lambda \sin \phi) = \pi W_\lambda \sin \phi$.

Both electric field vectors $E_V(t)$ and $E_H(t)$ are in the xy plane tangent to the circle of radius d, which is measured from the center of the circuit. Therefore, from Eqs. 3-11 and 3-15, the total radiated electric field from this circuit at observation points in the xy plane will be

$$E_{xy}(t) = E_V(t) + E_H(t)$$

$$= \left\{ \frac{Z_0|I|W_\lambda}{d} \cos \phi \sin[\pi(L_\lambda' + W_\lambda' - L_\lambda \cos \phi)] \right\} \sin\left[2\pi f \left(t - \frac{d}{c} \right) \right]$$

$$- \left\{ \frac{Z_0|I|W_\lambda}{d} \frac{\sin^2 \phi \sin[\pi L_\lambda(\cos \phi - c/v_p)]}{\cos \phi - c/v_p} \right\} \sin\left[2\pi f \left(t - \frac{d}{c} \right) \right]$$

$$= E_{xy}(\phi) \sin\left[2\pi f \left(t - \frac{d}{c} \right) \right] \qquad (3\text{-}16)$$

Thus,

$$E_{xy}(\phi) = \frac{Z_0|I|W_\lambda}{d} \cos \phi \sin[\pi(L_\lambda' + W_\lambda' - L_\lambda \cos \phi)]$$

$$- \frac{Z_0|I|W_\lambda}{d} \frac{\sin^2 \phi \sin[\pi L_\lambda(\cos \phi - c/v_p)]}{\cos \phi - c/v_p} \qquad (3\text{-}17)$$

$E_{xy}(\phi)$ is the xy-plane pattern factor of the radiated electric field of the circuit of Fig. 3-1, when its width is $W \le \frac{1}{16}(v_p/f) \le \frac{1}{16}(c/f)$. Again, unless it is stated otherwise, this width restriction will be assumed in all of the discussions that follow.

3.4.2   The xz-Plane Radiation Pattern

Consider the radiation of the circuit of Fig. 3-1 in the xz plane. In Fig. 3-7a the left-hand current element of length W is seen to be a distance $d + (L/2)\cos \xi$ from any observation point in the xz plane, and its current is $i[t + (L + W)/2v_p]$, which is positive in the positive y direction. Therefore, from Fig. 3-2, the positive direction of $E_s(t)$, the electric field it radiates, is seen to be the negative y direction. Also, $E_W(\xi)$ will be constant for all ξ, and when $\xi = \phi = 0$, then $E_W(\xi) = E_W(\phi) = -z_0|I|W_\lambda/2d$. Thus, in the xz plane the left-hand current element radiates

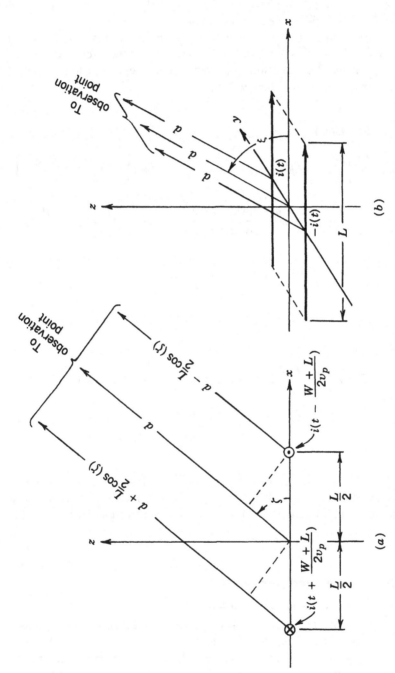

Figure 3-7. The geometry for finding the contribution of each of the current segment pairs to the *xz*-plane pattern factor of the radiated electric field of the circuit of Fig. 3-1: (*a*) source and load current segments; (*b*) connecting current segments.

$$E_s(t) = E_W(\xi) \cos\left\{2\pi f\left[t + \frac{L+W}{2v_p} - \frac{1}{c}\left(d + \frac{L}{2}\cos\xi\right)\right]\right\}$$

$$= \frac{-Z_0|I|W_\lambda}{2d}\cos\left[2\pi f\left(t - \frac{d}{c}\right) - \pi(L_\lambda\cos\xi - L'_\lambda - W'_\lambda)\right] \quad (3\text{-}18)$$

The right-hand current element of length W is a distance $d - (L/2)\cos\xi$ from any observation point in the xz plane, and its current is $i[t - (L+W)/2v_p]$. The right-hand current element and the left-hand current element have positive currents propagating in opposite directions. Therefore, since $W \leq \frac{1}{16}(v_p/f) \leq \frac{1}{16}(c/f)$, it also follows that $E_W(\xi - \pi) = -E_W(\xi)$, and the electric field radiated in the xz plane by the right-hand current element is

$$E_r(t) = E_W(\xi - \pi)\cos\left\{2\pi f\left[t - \frac{L+W}{2v_p} - \frac{1}{c}\left(d - \frac{L}{2}\cos\xi\right)\right]\right\}$$

$$= \frac{Z_0|I|W_\lambda}{2d}\cos\left[2\pi f\left(t - \frac{d}{c}\right) + \pi\left(L_\lambda\cos\xi - L'_\lambda - W'_\lambda\right)\right] \quad (3\text{-}19)$$

Once again using the identity $\cos(a - b) - \cos(a + b) = 2\sin(a)\sin(b)$, with $a = 2\pi f(t - d/c)$, and $b = \pi(L_\lambda\cos\xi - L'_\lambda - W'_\lambda)$, it follows from Eqs. 3-18 and 3-19 that

$$E_s(t) + E_r(t) = E_W(\xi)\cos\left[2\pi f\left(t - \frac{d}{c}\right) - \pi(L_\lambda\cos\xi - L'_\lambda - W'_\lambda)\right]$$

$$- E_W(\xi)\cos\left[2\pi f\left(t - \frac{d}{c}\right) + \pi(L_\lambda\cos\xi - L'_\lambda - W'_\lambda)\right]$$

$$= 2E_W(\xi)\sin[\pi(L_\lambda\cos\xi - L'_\lambda - W'_\lambda)]\sin\left[2\pi f\left(t - \frac{d}{c}\right)\right]$$

$$= \frac{Z_0|I|W_\lambda}{d}\sin[\pi(L'_\lambda + W'_\lambda - L_\lambda\cos\xi)]\sin\left[2\pi f\left(t - \frac{d}{c}\right)\right]$$

$$(3\text{-}20)$$

Note in Fig. 3-7b that the current segments of length L are parallel to the x axis and equidistant from all observation points in the xz plane. The currents in those segments are equal in magnitude and oppositely directed, but they propagate in the same direction. In addition, these segments are only a distance $W \leq \frac{1}{16}(c/f)$ apart. Therefore, for all practical purposes, at every observation point in the xz plane, the radiated electric fields of those current segments are equal and oppositely directed. In other words, at any observation point in the xz plane, the effective total contribution of those current segments to the radiated electric field is zero.

Because the segments of length L do not contribute, it follows from Eq. 3-20 that the total radiated electric field of this circuit in the xz plane is

$$E_{xz}(t) = E_s(t) + E_r(t)$$

$$= E_{xz}(\xi) \sin\left[2\pi f\left(t - \frac{d}{c}\right)\right]$$

where

$$E_{xz}(\xi) = \frac{Z_0|I|W_\lambda}{d} \sin[\pi(L'_\lambda + W'_\lambda - L_\lambda \cos \xi)] \tag{3-21}$$

This is the xz plane pattern factor of the electric field radiated by the circuit of Fig. 3-1, when $W_\lambda \leq \frac{1}{16}$.

3.4.3 The yz-Plane Radiation Pattern

Finally, referring to Fig. 3-8a, it is seen that in the yz plane the upper current segment of length L is a distance of $d - (W/2)\sin\theta$ from the observation point. And, referring again to Fig. 3-2, it is seen that in the yz plane the positive direction of the field of this segment will be the negative x direction, normal to the yz plane. Therefore, at observation points in the yz plane, that segment has a radiated electric field normal to the yz plane of

$$E_u(t) = E_L(\theta) \cos\left\{2\pi f\left[t - \frac{1}{c}\left(d - \frac{W}{2}\sin\theta\right)\right]\right\}$$

$$= E_L(\theta) \cos\left[2\pi f\left(t - \frac{d}{c}\right) + \pi W_\lambda \sin\theta\right] \tag{3-22}$$

The lower current segment of length L is a distance $d + (W/2)\sin\theta$ from the observation point and positive in the direction opposite to that of the upper segment, although both currents propagate in the same direction. Therefore, at observation points in the yz plane, the lower current segment of length L has a radiated electric field normal to the yz plane of

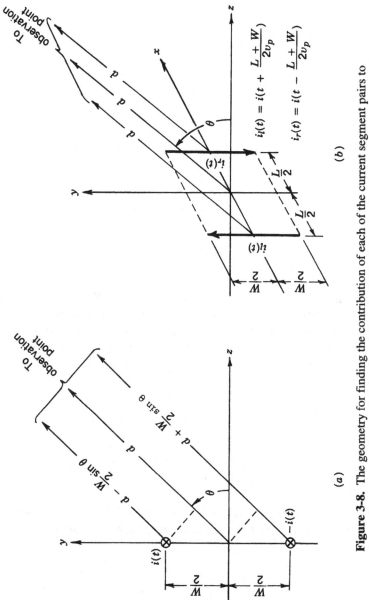

Figure 3-8. The geometry for finding the contribution of each of the current segment pairs to the yz-plane pattern factor of the radiated electric field of the circuit of Fig. 3-1: (a) connecting-current segments; (b) source and load current segments.

In figure (b):

$$i_l(t) = i\left(t + \frac{L+W}{2v_p}\right)$$

$$i_r(t) = i\left(t - \frac{L+W}{2v_p}\right)$$

$$E_\ell(t) = -E_L(\theta)\cos\left\{2\pi f\left[t - \frac{1}{c}\left(d + \frac{W}{2}\sin\theta\right)\right]\right\}$$

$$= -E_L(\theta)\cos\left[2\pi f\left(t - \frac{d}{c}\right) - \pi W_\lambda \sin\theta\right] \tag{3-23}$$

Thus, since $\cos(a + b) - \cos(a - b) = -2\sin(a)\sin(b)$, with $a = 2\pi f(t - d/c)$ and $b = \pi W_\lambda \sin\theta$, the sum of $E_u(t)$ and $E_\ell(t)$ is seen to be

$$E_u(t) + E_\ell(t) = E_L(\theta)\cos\left[2\pi f\left(t - \frac{d}{c}\right) + \pi W_\lambda \sin\theta\right]$$

$$- E_L(\theta)\cos\left[2\pi f\left(t - \frac{d}{c}\right) - \pi W_\lambda \sin\theta\right]$$

$$= -2E_L(\theta)\sin(\pi W_\lambda \sin\theta)\sin\left[2\pi f\left(t - \frac{d}{c}\right)\right] \tag{3-24}$$

As shown in Fig. 3-8b, the current elements of length W and parallel to the y axis are both a distance d from observation points in the yz plane. Therefore, the forward element on the negative x axis with current $i[t + (W + L)/2v_p]$ propagating in the positive y direction has a radiated electric field in the yz plane of

$$E_f(t) = E_W(\theta)\cos\left[2\pi f\left(t + \frac{W + L}{2v_p}\right) - \frac{d}{c}\right]$$

$$= E_W(\theta)\cos\left[2\pi f\left(t - \frac{d}{c}\right) + \pi(W_\lambda' + L_\lambda')\right] \tag{3-25}$$

The rear element with the positive current $i[t - (W + L)/2v_p]$ propagating in the negative y direction has a radiated electric field in the yz plane of

$$E_r(t) = E_W(\theta - \pi)\cos\left[2\pi f\left(t - \frac{W + L}{2v_p}\right) - \frac{d}{c}\right]$$

$$= -E_W(\theta)\cos\left[2\pi f\left(t - \frac{d}{c}\right) - \pi(W_\lambda' + L_\lambda')\right] \tag{3-26}$$

Thus,

$$E_f(t) + E_r(t) = E_W(\theta) \cos\left[2\pi f\left(t - \frac{d}{c}\right) + \pi(W'_\lambda + L'_\lambda)\right]$$

$$- E_W(\theta) \cos\left[2\pi f\left(t - \frac{d}{c}\right) - \pi(W'_\lambda + L'_\lambda)\right]$$

$$= -2E_W(\theta)\sin[\pi(W'_\lambda + L'_\lambda)]\sin\left[2\pi f\left(t - \frac{d}{c}\right)\right]$$

$$= \frac{-Z_0|I|W_\lambda}{d}\sin\theta\,\sin[\pi(W'_\lambda + L'_\lambda)]\sin\left[2\pi f\left(t - \frac{d}{c}\right)\right]$$

(3-27)

This result was obtained by once again using the identity $\cos(a+b) - \cos(a-b)$ $= -2\sin(a)\sin(b)$, with $a = 2\pi f(t - d/c)$ and $b = \pi(W'_\lambda + L'_\lambda)$.

Again referring to Fig. 3-2, it is seen that the direction of the electric field $E_u(t) + E_\ell(t)$ will be normal to the yz plane, whereas the direction of the electric field $E_f(t) + E_r(t)$ will lie in the yz plane. Thus, these two electric fields are perpendicular to one another. As a result, from Eqs. 3-24 and 3-27, it follows that the magnitude of their total radiated electric field will be

$$|E_{yz}(\theta)| = 2\sqrt{\{E_W(\theta)\sin[\pi(W'_\lambda + L'_\lambda)]\}^2 + [E_L(\theta)\sin(\pi W_\lambda \sin\theta)]^2} \quad (3\text{-}28)$$

This is the magnitude of the yz-plane pattern factor of the radiated electric field from the circuit of Fig. 3-1.

With these general expressions for their coordinate-plane pattern factors, or their magnitude, the coordinate-plane radiation patterns of numerous rectangular circuits can be found.

3.5   THE RADIATIONS OF SMALL CIRCUITS

The radiation patterns of small circuits are found by assuming all of the current segments to be current elements. Any rectangular circuit will be considered small when its length L and width W are both less than $\lambda/16$, where $\lambda = v_p/f$ is one wavelength of the current in the circuit. As noted earlier, v_p is the current's propagation velocity, f is its frequency, and $v_p \leq c$ the speed of light. In other words, in a small circuit, $L_\lambda = L(f/c) \leq L'_\lambda = L(f/v_p) \leq \frac{1}{16}$, and $W_\lambda = W(f/c) \leq W'_\lambda = W(f/v_p) \leq \frac{1}{16}$.

Now, suppose the current segments of the circuit of Fig. 3-1 are all current elements. Then the xy-plane pattern factor of the circuit can be found from the geometry of Figs. 3-5 and 3-6 and Eq. 3-17 as follows. Suppose, initially, that

$v_p = c$, which is the most it can be. Then, because the circuit is small

$$\frac{\pi}{16} \leq \pi(L'_\lambda + W'_\lambda - L_\lambda \cos \phi) \leq \frac{3\pi}{16}$$

and

$$-\frac{\pi}{8} \leq \pi L_\lambda \left(\cos \phi - \frac{c}{v_p} \right) = \pi L_\lambda (\cos \phi - 1) \leq 0 \qquad (3\text{-}29)$$

From these relationships it is seen that

$$\sin[\pi(L'_\lambda + W'_\lambda - L_\lambda \cos \phi)] \cong \pi(L'_\lambda + W'_\lambda - L_\lambda \cos \phi) \qquad (3\text{-}30)$$

and

$$\frac{\sin[\pi L_\lambda (\cos \phi - c/v_p)]}{\cos \phi - c/v_p} = \frac{\sin[\pi L_\lambda (\cos \phi - 1)]}{\cos \phi - 1}$$

$$\cong \pi L_\lambda \qquad (3\text{-}31)$$

Thus, Eqs. 3-17, 3-30 and 3-31 imply that a small circuit's xy-plane pattern factor is

$$E_{xy}(\phi) \cong \frac{Z_0 |I| W_\lambda}{d} \{[\pi(L'_\lambda + W'_\lambda - L_\lambda \cos \phi)] \cos \phi - \pi L_\lambda \sin^2 \phi\}$$

$$= \frac{\pi Z_0 |I| W_\lambda}{d} [(L'_\lambda + W'_\lambda) \cos \phi - L_\lambda \cos^2 \phi - L_\lambda \sin^2 \phi]$$

$$= \frac{\pi Z_0 |I| W_\lambda}{d} [(L'_\lambda + W'_\lambda) \cos \phi - L_\lambda]. \qquad (3\text{-}32)$$

The trigonometric identity $\sin^2 \phi + \cos^2 \phi = 1$ yields the final expression for $E_{xy}(\phi)$ in these equations.

The xz-plane and yz-plane radiation pattern factors for small circuits are found below. However, before proceeding, suppose that the phase differences in the current elements had not been considered in obtaining Eq. 3-32. If phase differences in the current around this circuit are assumed to be negligible, then that is the same as assuming propagation time $t_p \rightarrow 0$ and phase velocity $v_p \rightarrow \infty$. Thus, $t_p = (W + L)/2v_p \rightarrow 0$, and the currents in the current elements of length W, given in Eqs. 3-3 and 3-4, will now both be $i(t) = |I| \sin(2\pi f t)$, because $\pi(L'_\lambda + W'_\lambda) \rightarrow 0$. Thus, from Eq. 3-32, because $\pi(L'_\lambda + W'_\lambda) \rightarrow 0$, the magnitude of the xy-plane pattern factor becomes

$$|E'_{xy}(\phi)| = \frac{\pi Z_0 |I| L_\lambda W_\lambda}{d} \tag{3-33}$$

The prime (') is included in the notation for $E'_{xy}(\phi)$ here to indicate that no current phase difference is assumed.

The assumption that the current has the same phase everywhere in small circuits such as this, or in similarly small loop antennas, is very often made (e.g., Ott, 1988, p. 300; Weeks, 1968, p. 56; Kraus, 1988, p. 238; Terman, 1955, p. 907). That assumption is assumed to be valid, for example, when $L'_\lambda \leq \frac{1}{16}$ and $W'_\lambda \leq \frac{1}{16}$, so that the circuit, or small loop antenna, has a perimeter $2(L'_\lambda + W'_\lambda) \leq \frac{1}{4}$. The conclusion drawn is that in its own plane, a circuit or loop of that size or smaller will have a maximum radiated electric field that is closely equal to the expression given for $|E'_{xy}(\phi)|$ in Eq. 3-33.

However, suppose phase differences are *not* neglected in the current elements of this circuit. Since $v_p \leq c$, it is clear that $L_\lambda \leq L'_\lambda \leq \frac{1}{16}$ and $W_\lambda \leq W'_\lambda \leq \frac{1}{16}$. Therefore, assuming that $v_p = c$, which is the most v_p can be, then $L_\lambda = L'_\lambda = W_\lambda = W'_\lambda$, and it follows from Eq. 3-32 that

$$|E_{xy}(\phi)| = \frac{\pi Z_0 |I| W_\lambda}{d} |L_\lambda - (L_\lambda + W_\lambda) \cos \phi| \tag{3-34}$$

From Eqs. 3-33 and 3-34, it can readily be seen that if the current element lengths are $L_\lambda = W_\lambda = \frac{1}{16}$ and $\phi = \pi$, for example, making $\cos \phi = -1$, then

$$\frac{|E_{xy}(\phi)|}{|E'_{xy}(\phi)|} = \frac{|2L_\lambda + W_\lambda|}{L_\lambda}$$

$$= 3 \tag{3-35}$$

In other words, the maximum radiated electric field of this circuit in the xy plane is *three times* that predicted when phase differences are neglected.

A graphical comparison of the xy-plane radiation patterns $|E_{xy}(\phi)|$ and $|E'_{xy}(\phi)|$ for *all* values of ϕ is given in Fig. 3-9. It is assumed there that $L_\lambda = L'_\lambda = W_\lambda = W'_\lambda = \frac{1}{16}$. Based on that assumption, the expressions for $|E_{xy}(\phi)|$ and $|E'_{xy}(\phi)|$ obtained from Eqs. 3-33 and 3-34 are

$$|E'_{xy}(\phi)| = \frac{\pi Z_0 |I|}{256d}$$

and

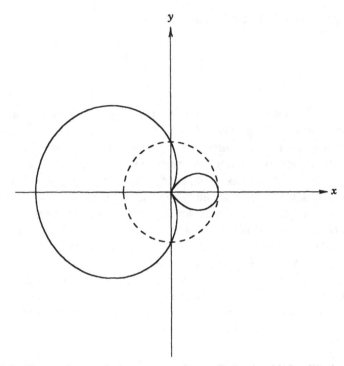

Figure 3-9. The xy-plane radiation pattern of a small circuit with $L = W$ when current-segment phase differences have been assumed (——) and when no phase differences have been assumed (---).

$$|E_{xy}(\phi)| = \frac{\pi Z_0 |I|}{256d} \, |1 - 2\cos\phi| \qquad (3\text{-}36)$$

It should be quite clear from the above and from Fig. 3-9 that phase differences should not be neglected when predicting the radiations of even the smallest circuit currents. It can be seen in Fig. 3-9 that $|E_{xy}(\phi)| > |E'_{xy}(\phi)|$ everywhere in the left half of the xy plane.

Now consider the xz-plane radiation geometry of Fig. 3-7 when $L_\lambda \le L'_\lambda \le \frac{1}{16}$ and $W_\lambda \le W'_\lambda \le \frac{1}{16}$. The pattern factors of both current elements of length W in the xz plane, their plane of maximum radiation, is

$$E_W(\xi) = \frac{Z_0 |I| W_\lambda}{2d} \qquad (3\text{-}37)$$

The pattern factors of the current elements of length L are irrelevant, because, as previously noted, for all practical purposes their radiations cancel in the xz plane. Thus, from Eqs. 3-21 and 3-37, it is seen that the xz-plane pattern factor of this circuit current is

$$E_{xz}(\xi) = \frac{Z_0|I|W_\lambda}{d} \sin[\pi(L_\lambda \cos \xi - L'_\lambda - W'_\lambda)]$$

$$\cong \frac{\pi Z_0|I|W_\lambda}{d} (L_\lambda \cos \xi - L'_\lambda - W'_\lambda) \qquad (3\text{-}38)$$

Notice that if phase differences are neglected here, then $(L'_\lambda + W'_\lambda) \rightarrow 0$, and $E_{xz}(\xi)$ becomes

$$E'_{xz}(\xi) \cong \frac{\pi Z_0|I|L_\lambda W_\lambda}{d} \cos \xi \qquad (3\text{-}39)$$

From Eqs. 3-38 and 3-39, if $L_\lambda = L'_\lambda = W_\lambda = W'_\lambda = \frac{1}{16}$, and $\xi = \pi$, then that makes $\cos \xi = -1$, and

$$\frac{|E_{xz}(\xi)|}{|E'_{xz}(\xi)|} \cong \frac{|L_\lambda \cos \xi - L_\lambda - W_\lambda|}{|L_\lambda \cos \xi|}$$

$$= \frac{2L_\lambda + W_\lambda}{L_\lambda}$$

$$= 3 \qquad (3\text{-}40)$$

The graphical comparison of these radiation patterns in Fig. 3-10 shows that

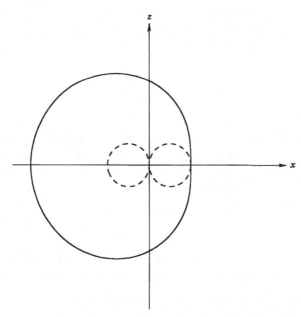

Figure 3-10. The xz-plane radiation pattern of a small circuit with $L = W$ when current-segment phase differences have been assumed (—), and when no phase differences have been assumed (---).

$|E_{xz}(\xi)| > |E'_{xz}(\xi)|$ for all ξ except $\xi = 0$. It is again assumed there that $L_\lambda = L'_\lambda = W_\lambda = W'_\lambda = \frac{1}{16}$, and the resulting expressions for $|E_{xz}(\xi)|$ and $|E'_{xz}(\xi)|$, obtained from Eqs. 3-38 and 3-39, are

$$|E_{xz}(\xi)| = \frac{Z_0|I|}{256d} \, |\cos \xi - 2|$$

and

$$|E'_{xz}(\xi)| = \frac{Z_0|I|}{256d} \, |\cos \xi| \qquad (3\text{-}41)$$

Finally, the radiation pattern of the small circuit of Fig. 3-1 is found in the yz plane. As shown in Fig. 3-8, the current elements of length L are both normal to and centered on the yz plane, which is their plane of maximum radiation. Therefore, the yz-plane pattern factor of each is

$$E_L(\theta) = \frac{Z_0|I||L_\lambda}{2d} \qquad (3\text{-}42)$$

The current elements of length W are both parallel to the y axis, and the angle θ is measured from the z axis. As a result, the current element that has the source centered in it, in which positive current propagates upward, has the yz-plane pattern factor

$$E_W(\theta) = \frac{Z_0|I||W_\lambda}{2d} \, \cos \theta \qquad (3\text{-}43)$$

And the pattern factor of the current element with the load in it, in which positive current propagates downward, is the negative of that of Eq. 3-43.

Therefore, from Eqs. 3-28, 3-42, and 3-43, it follows that whenever $L_\lambda \leq L'_\lambda \leq \frac{1}{16}$ and $W_\lambda \leq W'_\lambda \leq \frac{1}{16}$, the circuit of Fig. 3-1 has a yz-plane pattern factor with the magnitude

$$
\begin{aligned}
|E_{yz}(\theta)| &= \frac{Z_0|I|}{d} \sqrt{\{W_\lambda \cos \theta \sin[\pi(W'_\lambda + L'_\lambda)]\}^2 + [L_\lambda \sin(\pi W_\lambda \sin \theta)]^2} \\
&\cong \frac{Z_0|I|}{d} \sqrt{[\pi W_\lambda(W'_\lambda + L'_\lambda) \cos \theta]^2 + (\pi L_\lambda W_\lambda \sin \theta)^2} \\
&= \frac{\pi Z_0|I||W_\lambda}{d} \sqrt{(W'_\lambda + L'_\lambda)^2 \cos^2 \theta + L_\lambda^2 \sin^2 \theta} \qquad (3\text{-}44)
\end{aligned}
$$

In this case, if the phase differences in the circuit's current elements are assumed to be negligible, then $(W'_\lambda + L'_\lambda) \rightarrow 0$, so that

$$|E'_{yz}(\theta)| = \frac{\pi Z_0 |I| L_\lambda W_\lambda}{d} |\sin \theta| \tag{3-45}$$

Thus, in the yz plane,

$$\frac{|E_{yz}(\theta)|}{|E'_{yz}(\theta)|} = \frac{\sqrt{(W'_\lambda + L'_\lambda)^2 \cos^2 \theta + L_\lambda \sin^2 \theta}}{L_\lambda |\sin \theta|} \tag{3-46}$$

A graphical comparison of the radiation patterns of $|E_{yz}(\theta)|$, and $|E'_{yz}(\theta)|$, with $L_\lambda = L'_\lambda = W_\lambda = W'_\lambda = \frac{1}{16}$, is made in Fig. 3-11 where

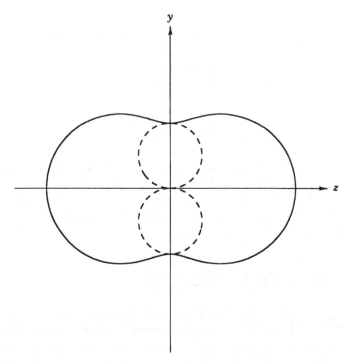

Figure 3-11. The yz-plane radiation pattern of a small circuit with $L = W$ when current-segment phase differences have been assumed (—), and when no phase differences have been assumed (---).

$$|E_{yz}(\theta)| = \frac{\pi Z_0 |I|}{16d} \sqrt{\frac{\cos^2 \theta}{64} + \frac{\sin^2 \theta}{256}}$$

$$= \frac{\pi Z_0 |I|}{256d} \sqrt{4\cos^2 \theta + \sin^2 \theta}$$

and

$$|E'_{yz}(\theta)| = \frac{\pi Z_0 |I|}{256d} |\sin \theta| \qquad (3\text{-}47)$$

and it is seen that $|E_{yz}(\theta)| > |E'_{yz}(\theta)|$, except when $\theta = \pm\pi/2$.

The maximum radiation from the small circuit considered here is seen to occur in the negative x direction. In general, the maximum radiation of small rectangular circuits will be in the direction from the halfway point to the source. The mathematical expression for the maximum electric field can be obtained either from $|E_{xy}(\phi)|$ by setting $\phi = \pi$ or from $|E_{xz}(\xi)|$ by setting $\xi = \pi$. The result will be the same in either case. Using the latter with $\xi = \pi$, it follows from Eq. 3-38 that the maximum radiated electric field is

$$\max |E| = \frac{\pi Z_0 |I| W_\lambda}{d} |L_\lambda \cos(\pi) - (L_\lambda + W_\lambda)|$$

$$= \frac{\pi Z_0 |I| W_\lambda}{d} (2L_\lambda + W_\lambda)$$

$$= \frac{\pi Z_0}{c^2 d} |I| f^2 W (2L + W) \qquad (3\text{-}48)$$

This expression gives a close approximation of the maximum radiated electric field of a small rectangular circuit in free space. The circuit is small, because $W_\lambda = Wf/c \le \frac{1}{16}$ and $L_\lambda = Lf/c \le \frac{1}{16}$, which makes it reasonable to assume that $\pi(2L_\lambda + W_\lambda) \cong \sin[\pi(2L_\lambda + W_\lambda)]$. It has also been assumed here, as it will be hereafter, that current phase velocity is $v_p = c$.

3.6   THE RADIATIONS OF LONGER CIRCUITS

The coordinate-plane radiation patterns of circuit currents that have connecting current segments longer than current elements can easily be obtained. General expressions for the pattern factors of circuit currents, or their magnitudes, obtained from Eqs. 3-17, 3-21, and 3-28 are given below, and it is assumed, as noted above, that $v_p = c$, so that $L'_\lambda = L_\lambda$ and $W'_\lambda = W_\lambda$. Therefore, the xy-plane radiation pattern factor of a rectangular circuit current of any length L_λ and width $W_\lambda \le \frac{1}{16}$, such as that of the circuit of Fig. 3-1, is

$$E_{xy}(\phi) = 2\{E_W(\phi) \sin[\pi L_\lambda(\cos \phi - 1) - \pi W_\lambda] - \pi W_\lambda E_L(\phi) \sin \phi\} \qquad (3\text{-}49)$$

The xz-plane pattern factor of that circuit current is

$$E_{xz}(\xi) = 2E_W(\xi) \sin[\pi L_\lambda(\cos \phi - 1) - \pi W_\lambda] \qquad (3\text{-}50)$$

And, the magnitude of its yz-plane pattern factor is

$$|E_{yz}(\theta)| = 2\sqrt{\{E_W(\theta) \sin[\pi(L_\lambda + W_\lambda)]\}^2 + [\pi W_\lambda E_L(\theta) \sin \theta]^2}. \qquad (3\text{-}51)$$

Several graphic examples derived from these expressions are now given.

Example 3-1   Suppose the dimensions of the circuit of Fig. 3-1 are $L_\lambda \ge L_{e\lambda}$ and $W_\lambda = W_{e\lambda}$. The current elements of length W are both parallel to the y axis and centered on the x axis. Therefore, in Eq. 3-49

$$E_W(\phi) = \frac{-Z_0|I|W_{e\lambda}}{2d} \cos \phi \qquad (3\text{-}52)$$

The current segments of length L_λ are both parallel to the x axis and centered on the y axis, so that, assuming $v_p = c$, in Eq. 3-49

$$\begin{aligned} E_L(\phi) &= \frac{Z_0|I|}{2\pi d} \sin \phi \; \frac{\sin[\pi L_\lambda(\cos \phi - 1)]}{\cos \phi - 1} \\[2mm] &= -\frac{Z_0|I|}{2\pi d} (\cos \phi + 1) \; \frac{\sin[\pi L_\lambda(\cos \phi - 1)]}{\sin \phi} \end{aligned} \qquad (3\text{-}53)$$

The latter expression follows from the trigonometric identity $\sin^2 \phi + \cos^2 \phi = 1$, and the relationship $\cos^2 \phi - 1 = (\cos \phi + 1)(\cos \phi - 1)$.

Thus, from Eqs. 3-49, 3-52, and 3-53, it follows that the xy-plane radiation pattern factor of this circuit current is

$$E_{xy}(\phi) = 2\left\{ \frac{-Z_0|I|W_{e\lambda}}{2d} \cos\phi\sin[\pi L_\lambda(\cos\phi - 1) - \pi W_{e\lambda}]\right.$$

$$\left. - \pi W_{e\lambda}E_L(\phi)\sin\phi \right\}$$

$$= \frac{-Z_0|I|W_{e\lambda}}{d}\{\cos\phi\sin[\pi L_\lambda(\cos\phi - 1) - \pi W_{e\lambda}] - (\cos\phi + 1)$$

$$\cdot \sin[\pi L_\lambda(\cos\phi - 1)]\} \tag{3-54}$$

The xy-plane radiation patterns obtained with $|E_{xy}(\phi)|$, when $L_\lambda = \frac{1}{2}$, 1, and 2, and $W_{e\lambda} = \frac{1}{16}$, are illustrated in Fig. 3-12.

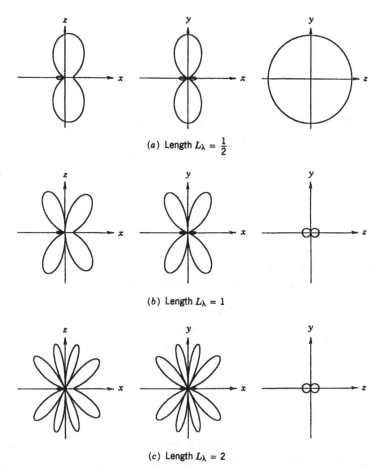

(a) Length $L_\lambda = \frac{1}{2}$

(b) Length $L_\lambda = 1$

(c) Length $L_\lambda = 2$

Figure 3-12. The radiation patterns of rectangular circuits of width $W_\lambda = \frac{1}{16}$ and the lengths indicated, when each has the same sinusoidal current.

The xz plane is the plane of maximum radiation of the current elements of length $W_\lambda = W_{e\lambda}$. Therefore, in Eq. 3-50

$$E_W(\xi) = \frac{Z_0|I|W_{e\lambda}}{2d} \qquad (3\text{-}55)$$

As previously noted, the fields of the current segments of length L_λ cancel everywhere in the xz plane. Therefore, with $W_\lambda = W_{e\lambda} \leq \frac{1}{16}$, Eqs. 3-50 and 3-55 imply that

$$E_{xz}(\xi) = 2E_W(\xi)\sin(\pi L_\lambda(\cos\xi - 1) - \pi W_{e\lambda})$$

$$= \frac{Z_0|I|W_{e\lambda}}{d}\ \sin(\pi L_\lambda(\cos\xi - 1) - \pi W_{e\lambda}) \qquad (3\text{-}56)$$

The xz-plane radiation patterns of this circuit current, obtained with $|E_{xz}(\xi)|$, when $L_\lambda = \frac{1}{2}$, 1, and 2, and $W_{e\lambda} = \frac{1}{16}$, are illustrated in Fig. 3-12.

Finally, the current segments of length L are both normal to the yz plane. Therefore, each has a constant pattern factor in the yz plane that is equal to either their xy-plane pattern factor when $\phi = \pi/2$ or their xz-plane pattern factor when $\xi = \pi/2$. Thus, it follows from Eq. 3-53, for $E_L(\phi)$, with $\phi = \pi/2$, that the expression for $E_L(\theta)$ in Eq. 3-51 is

$$E_L(\theta) \cong \frac{Z_0|I|}{2\pi d}\ \sin(\pi L_\lambda) \qquad (3\text{-}57)$$

The current elements of length $W_\lambda = W_{e\lambda}$, which are parallel to the y axis, both have a yz-plane pattern factor with the magnitude

$$|E_W(\theta)| = \frac{Z_0|I|W_{e\lambda}}{2d}\ |\cos\theta| \qquad (3\text{-}58)$$

Therefore, with $W_\lambda = W_{e\lambda} \leq \frac{1}{16}$ and Eqs. 3-51, 3-57, and 3-58, it follows that

$$|E_{yz}(\theta)| = 2\sqrt{\{E_W(\theta)\sin[\pi(W_{e\lambda} + L_\lambda)]\}^2 + [\pi W_{e\lambda} E_L(\theta)\sin\theta]^2}$$

$$= 2\sqrt{\left\{\frac{Z_0|I|W_{e\lambda}}{2d}\ \cos\theta\sin[\pi(W_{e\lambda} + L_\lambda)]\right\}^2 + \left[\frac{Z_0|I|W_{e\lambda}}{2d}\ \sin(\pi L_\lambda)\sin\theta\right]^2}$$

$$= \frac{Z_0|I|W_{e\lambda}}{d}\ \sqrt{\cos^2\theta\sin^2[\pi(W_{e\lambda} + L_\lambda)] + \sin^2(\pi L_\lambda)\sin^2\theta} \qquad (3\text{-}59)$$

Radiation patterns in the yz plane obtained with Eq. 3-59, when $W_{e\lambda} = \frac{1}{16}$ and $L_\lambda = \frac{1}{2}$, 1, and 2, are illustrated in Fig. 3-12.

Although their directions change, the maximum magnitudes of the radiations in each of the patterns shown in Fig. 3-12 appear to be the same, although the length is twice doubled. To examine this more thoroughly, the maximum radiated electric fields of long, narrow, rectangular circuit currents in each of the coordinate planes will be found.

Since $|\sin[\pi L_\lambda(\cos \xi - 1) - \pi W_{e\lambda}]| \leq 1$, for any length L_λ, it is clear from Eq. 3-56 that

$$|E_{xz}(\xi)| \leq \frac{Z_0|I|W_{e\lambda}}{d} = \max |E_{xz}(\xi)| \qquad (3\text{-}60)$$

And since $\{\cos \theta \sin[\pi(W_{e\lambda} + L_\lambda)]\}^2 + [\sin \theta \sin(\pi L_\lambda)]^2 < \cos^2 \theta + \sin^2 \theta = 1$ for any length L_λ, it is clear from Eq. 3-59 that

$$|E_{yz}(\theta)| < \frac{Z_0|I|W_{e\lambda}}{d} = \max |E_{yz}(\theta)| \qquad (3\text{-}61)$$

The equal sign is not included here, because so long as $0 < W_{e\lambda} \leq \frac{1}{16}$, $\sin(\pi L_\lambda)$ and $\sin[\pi(W_{e\lambda} + L_\lambda)]$ cannot both equal 1 for any value of L_λ.

Finally, to find $\max |E_{xy}(\phi)|$, let $a = \pi L_\lambda(\cos \phi - 1)$ and $b = \pi W_{e\lambda}$ and recall that $\sin(a - b) = \sin(a)\cos(b) - \sin(b)\cos(a)$ for any a and b. Also note that if $W_{e\lambda} \leq \frac{1}{16}$, then $\cos(\pi W_{e\lambda}) \cong 1$ and $\sin(\pi W_{e\lambda}) \cong \pi W_{e\lambda}$. Thus,

$$\sin(\pi L_\lambda(\cos \theta - 1) - \pi W_{e\lambda})$$

$$= \sin[\pi L_\lambda(\cos \phi - 1)]\cos(\pi W_{e\lambda}) - \sin(\pi W_{e\lambda})\cos[\pi L_\lambda(\cos \phi - 1)]$$

$$\cong \sin[\pi L_\lambda(\cos \phi - 1)] - \pi W_{e\lambda}\cos[\pi L_\lambda(\cos \phi - 1)] \qquad (3\text{-}62)$$

From A-26 of Appendix A, $|A\sin(x) + B\cos(x)| \leq \sqrt{A^2 + B^2}$ for any A, B, and x. Thus, if $A = -1, B = -(\pi W_{e\lambda}\cos \phi)$, and $x = \pi L_\lambda(\cos \phi - 1)$, it follows that

$$|\cos \phi \sin[\pi L_\lambda(\cos \phi - 1) - \pi W_{e\lambda}] - (\cos \phi + 1)\sin[\pi L_\lambda(\cos \phi - 1)]|$$

$$\cong \left| \begin{array}{c} \cos \phi\{\sin[\pi L_\lambda(\cos \phi - 1)] - \pi W_{e\lambda}\cos[\pi L_\lambda(\cos \phi - 1)]\} \\ -(\cos \phi + 1)\sin[\pi L_\lambda(\cos \phi - 1)] \end{array} \right|$$

$$= |-\sin[\pi L_\lambda(\cos \phi - 1)] - (\pi W_{e\lambda}\cos \phi)\cos[\pi L_\lambda(\cos \phi - 1)]|$$

$$\leq \sqrt{(-1)^2 + (-\pi W_{e\lambda}\cos \phi)^2}$$

$$\leq \sqrt{1 + (\pi W_{e\lambda})^2}$$

$$\leq \sqrt{1 + (\pi/16)^2}$$

$$\leq 1.02 \tag{3-63}$$

Based on this result, it follows from Eq. 3-54 that for any length L_λ and for $W_{e\lambda} \leq \frac{1}{16}$,

$$|E_{xy}(\phi)| \leq 1.02 \, \frac{Z_0|I|W_{e\lambda}}{d} = \max |E_{xy}(\phi)| \tag{3-64}$$

Thus, the maximum radiated electric field of a long, narrow, rectangular circuit current in any of the three coordinate planes will be

$$\max |E| = 1.02 \, \frac{Z_0|I|W_{e\lambda}}{d} \tag{3-65}$$

From this it is clear that the maximum magnitudes of the radiated electric fields of the currents of long, narrow rectangular circuits are directly proportional to the widths of those circuits. Furthermore, those maxima are completely independent of the lengths of the circuits.

Example 3-2   The above observations are further illustrated in Figs. 3-13 and 3-14. The radiation patterns of the same sinusoidal current in circuits of lengths $L_\lambda = \frac{1}{4}, \frac{1}{2}$, and 1 and 2, 4, and 8 are each compared to a circle of radius $Z_0|I|W_\lambda/d = Z_0|I|/16d$, which equals the expected maximum. Thus, the radiations reach their expected maximums both in the xy plane and in the xz plane in every case. Doubling the length of the segments several times has no effect on the magnitude of the maximum attained.

The additional radiation patterns of the same sinusoidal current in circuits of lengths $L_\lambda = \frac{5}{8}, \frac{3}{4}, \frac{7}{8}, \frac{9}{8}, \frac{5}{4}$, and $\frac{11}{8}$ are shown in Figs. 3-15 and 3-16. In each of those figures the patterns are also compared to a circle of radius $Z_0|I|W_\lambda/d = Z_0|I|/16d$. In all of these figures, maximum radiations always occur in both the xy plane and the xz plane, but not necessarily in the yz plane. However, the directions of the maximum radiations vary, as noted previously, and the number of the maximums increases as the length of the circuit increases.

3.7   THE RADIATIONS OF MEDIUM-LENGTH CIRCUITS

It was seen in Eq. 3-48 that the maximum radiated electric field of a small circuit, for which $W_\lambda = W_{e\lambda} \leq \frac{1}{16}$ and $L_\lambda = L_{e\lambda} \leq \frac{1}{16}$, is

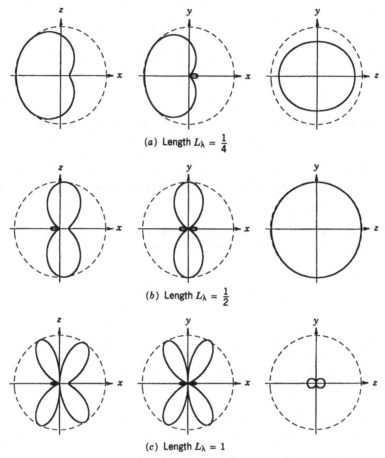

Figure 3-13. The radiation patterns of rectangular circuits of the lengths shown and the width $W_\lambda = \frac{1}{16}$, when each has the same sinusoidal current and each is compared to a circle of radius $Z_0|I|W_\lambda/d$.

$$\max |E| \cong \frac{\pi Z_0 |I| W_\lambda}{d} (2L_\lambda + W_\lambda) \tag{3-66}$$

And it was seen in the previous section that when $L_\lambda \geq \frac{1}{4}$ and $W_\lambda = W_{e\lambda} \leq \frac{1}{16}$, the maximum radiated electric field of a long rectangular circuit is

$$\max |E| \cong \frac{Z_0 |I| W_\lambda}{d} \tag{3-67}$$

It remains to determine what the maximum radiation of a narrow, rectangular circuit current will be when $W_\lambda = W_{e\lambda} \leq \frac{1}{16}$ and $\frac{1}{16} \leq L_\lambda < \frac{1}{4}$. Circuits in this category will be referred to here as narrow, medium-length circuits.

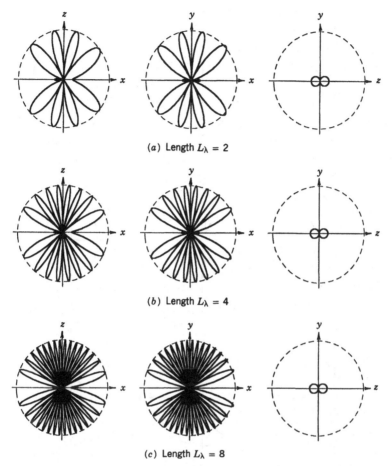

(a) Length $L_\lambda = 2$

(b) Length $L_\lambda = 4$

(c) Length $L_\lambda = 8$

Figure 3-14. The radiation patterns of rectangular circuits of the lengths shown and the width $W_\lambda = \frac{1}{16}$, when each has the same sinusoidal current and each is compared to a circle of radius $Z_0|I|W_\lambda/d$.

To make this determination it is necessary only to refer back to Eqs. 3-38. There the maximum radiated electric field of a small circuit was determined by noting that $W_\lambda \leq \frac{1}{16}$ and $L_\lambda \leq \frac{1}{16}$, so that $\sin[\pi(2L_\lambda + W_\lambda)] \cong \pi(2L_\lambda + W_\lambda)$. However, if $L_\lambda > \frac{1}{16}$, then

$$\max|E| = \frac{Z_0|I|W_\lambda}{d} \sin[\pi(2L_\lambda + W_\lambda)] \qquad (3\text{-}68)$$

And when $L_\lambda = \frac{1}{4} - W_\lambda/2$, then $\sin[\pi(2L_\lambda + W_\lambda)] = \sin[\pi(\frac{1}{2} - W_\lambda + W_\lambda)]$ $= \sin(\pi/2) = 1$ and $\max|E| = Z_0|I|W_\lambda/d$, so that Eq. 3-67 for long circuits is satisfied. Thus, Eq. 3-68 describes the maximum radiated electric field of any medium-length rectangular circuit current for which $W_\lambda \leq \frac{1}{16}$ and $\frac{1}{16} \leq$

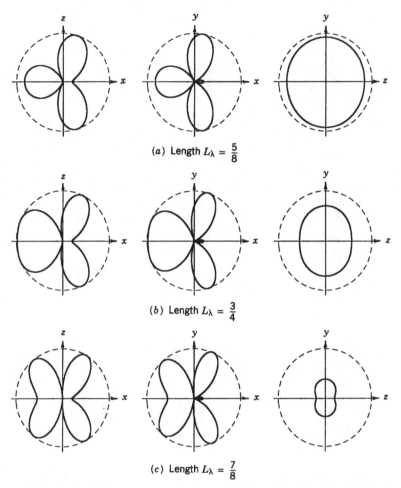

(a) Length $L_\lambda = \frac{5}{8}$

(b) Length $L_\lambda = \frac{3}{4}$

(c) Length $L_\lambda = \frac{7}{8}$

Figure 3-15. The radiation patterns of rectangular circuits of the lengths shown and the width $W_\lambda = \frac{1}{16}$, when each has the same sinusoidal current and each is compared to a circle of radius $Z_0|I|W_\lambda/d$.

$L_\lambda \leq \frac{1}{4} - W_\lambda/2 = \frac{7}{32}$. This effectively fills the gap between Eq. 3-66 for small circuits and Eq. 3-67 for long circuits, when the latter are defined for lengths $L_\lambda > \frac{1}{4} - W/2$.

Example 3-3   The radiation patterns of the same sinusoidal current in circuits of lengths from $L_\lambda = \frac{1}{16}$ to $L_\lambda = \frac{7}{32}$ are shown in Figs. 3-17 and 3-18, where they are each compared to a circle of radius $Z_0|I|W_\lambda/d = Z_0|I|/16d$. Here it is seen that the maximum radiation from medium-length circuits will also be directed from the halfway point to the source, as it is for small circuits.

It is also interesting to note that in all of the above examples, when L_λ is an integer multiple of $\frac{1}{2}$, the xy-plane radiation pattern is symmetric about the

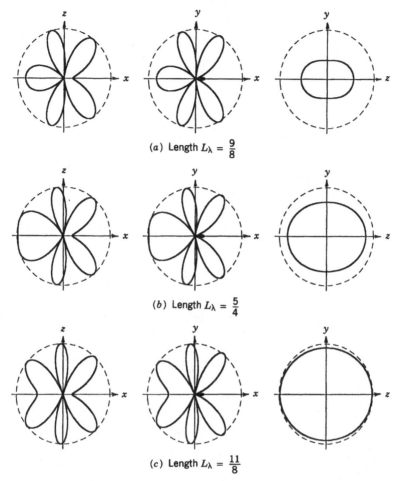

(a) Length $L_\lambda = \frac{9}{8}$

(b) Length $L_\lambda = \frac{5}{4}$

(c) Length $L_\lambda = \frac{11}{8}$

Figure 3-16. The radiation patterns of rectangular circuits of the lengths shown and the width $W_\lambda = \frac{1}{16}$, when each has the same sinusoidal current and each is compared to a circle of radius $Z_0|I|W_\lambda/d$.

origin. And when $n/2 < L_\lambda < (n + 1)/2$, two additional nodes of maximum radiation are formed, beyond those that occur for $L_\lambda = n/2$.

3.8 SUMMARY AND CONCLUSIONS

The radiation patterns and the maximum radiations of numerous rectangular circuit currents of different sizes were found here. Sinusoidal currents $i(t) = |I|\sin(2\pi f t)$ in small circuits, those with widths $W_\lambda = Wf/c \leq \frac{1}{16}$ and lengths $L_\lambda = Lf/c \leq \frac{1}{16}$, were found to have maximum radiated electric fields equal to

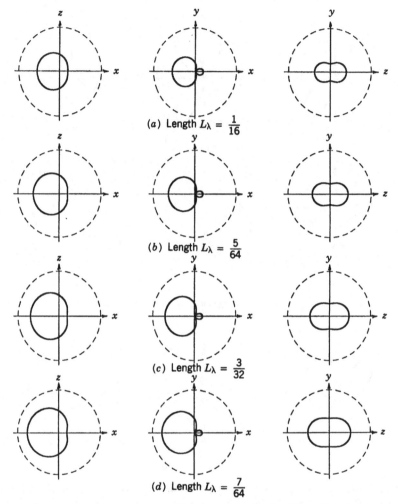

Figure 3-17. The radiation patterns of rectangular circuits of the lengths shown and the width $W_\lambda = \frac{1}{16}$, when each has the same sinusoidal current and each is compared to a circle of radius $Z_0|I|W_\lambda/d$.

$$\max |E| = \frac{\pi Z_0 |I| W_\lambda}{d} (2L_\lambda + W_\lambda) \tag{3-69}$$

This equation is valid up to the highest frequency f for which both $L_\lambda = Lf/c \le \frac{1}{16}$ and $W_\lambda = Wf/c \le \frac{1}{16}$. In these discussions, $c = 3 \times 10^8$ meters/second (the speed of light) is the assumed velocity of current propagation, so that c/f meters is one wavelength of a current of frequency f. Thus, the parameters L_λ and W_λ have no units, but they are the length and width of a circuit in wavelengths, or, more commonly, fractions of a wavelength. It was also seen, for small circuits,

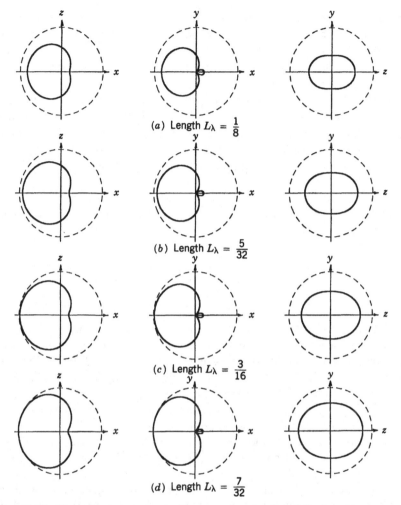

Figure 3-18. The radiation patterns of rectangular circuits of the lengths shown and the width $W_\lambda = \frac{1}{16}$, when each has the same sinusoidal current and each is compared to a circle of radius $Z_0|I|W_\lambda/d$.

that the direction of $\max|E|$ is the same as the direction from the circuit's load to its source, when they are at opposite ends of a rectangular circuit.

The same currents, $i(t) = |I|\sin(2\pi f t)$, in medium-length circuits, those with widths $W_\lambda \le \frac{1}{16}$ and lengths such that $\frac{1}{16} < L_\lambda \le \frac{1}{4} - W_\lambda/2$, were found to cause maximum radiated electric fields equal to

$$\max|E| = \frac{Z_0|I|W_\lambda}{d}\, \sin[\pi(2L_\lambda + W_\lambda)] \qquad (3\text{-}70)$$

This equation is valid for sinusoidal currents in rectangular circuits at all frequencies at which the above length and width requirements for medium-length circuits are both met. The direction of $\max|E|$ for medium-length circuits was also seen to be the direction from the circuit's load to its source when they are at opposite ends of the circuit.

Note that Eq. 3-70 describes $\max|E|$ for both small and medium-length circuits, based on the approximation $\sin[\pi(2L_\lambda + W_\lambda)] \cong \pi(2L_\lambda + W_\lambda)$, when $W_\lambda \leq \frac{1}{16}$ and $L_\lambda \leq \frac{1}{16}$. Therefore, when $W_\lambda \leq \frac{1}{16}$ and $L_\lambda \leq \frac{1}{4} - W_\lambda/2$, a circuit will be referred to as small or medium-length and Eq. 3-70 will be applicable. When both $W_\lambda \leq \frac{1}{16}$ and $L_\lambda \leq \frac{1}{16}$, a circuit will be small and either Eq. 3-69 or 3-70 will be applicable.

The same sinusoidal currents, $i(t) = |I| \sin(2\pi f t)$, in long circuits, those with widths $W_\lambda \leq \frac{1}{16}$ and lengths $L_\lambda > \frac{1}{4} - W_\lambda/2$, were found to cause maximum radiated electric fields

$$\max |E| = \frac{Z_0|I|W_\lambda}{d} \tag{3-71}$$

This equation is valid for sinusoidal currents in rectangular circuits at all frequencies at which the length and width requirements for long circuits are both met. In long circuits, maximum radiation was seen to occur in many different directions.

It is clear from Eqs. 3-69, 3-70, and 3-71 that reducing the width of a rectangular circuit will be far more effective in reducing the radiations of its current than reducing its length will be. In small circuits, for example, if $L = W$ and L is reduced to $L/2$, then $W(2L+W)$ is reduced from $3W^2$ to $2W^2$, which reduces $\max|E|$ to 67% of its previous value. If, instead, W is reduced to $W/2$, then $W(2L + W)$ is reduced from $3W^2$ to $1\frac{1}{4}W^2$, and $\max|E|$ is thereby reduced to 42% of its previous value.

On the other hand, reducing the length of a long rectangular circuit will have no effect on the magnitude of its maximum radiation, yet reducing its width will cause a proportional reduction in that magnitude. For long circuits, $\max|E|$ is entirely independent of the length of the circuit. For example, it is seen in Figs. 3-13 and 3-14 that doubling the length of rectangular circuits of lengths $L_\lambda \geq \frac{1}{4}$ changes the number and directions of their maximum radiations, but causes no change in their magnitude. Therefore, a reduction of L_λ in those circuits will be completely ineffective in reducing their radiations, unless it is sufficient to make $L_\lambda < \frac{1}{4}$.

Perhaps the most telling observation made in this chapter is that illustrated in Fig. 3-4. If a circuit's width W is made to approach zero, then the circuit current's radiation will also be made to approach zero. However, if the circuit's length L is made to approach zero, then the circuit current distribution, and its radiation, will be made to approach that of a dipole antenna. In other words, a rectangular circuit's capability to radiate may be enhanced, rather than reduced, if L is made to approach zero.

From a practical point of view, all circuits will not be planar and rectangular like those considered here. However, many circuit-current radiations can still be closely approximated by viewing circuits as two or more rectangular circuits, even when they are not necessarily in the same plane. Therefore, the rather restricted viewpoint taken in this chapter will be more generally valuable than it might first seem. This is discussed at greater length in Chapter 6.

The primary conclusion to be drawn from this chapter is the following. To reduce circuit-current radiations, the connecting currents from source to load should always have paths as close to one another as it is possible to make them, immediately upon leaving the source.

CHAPTER 4

FREQUENCY DOMAIN DESCRIPTIONS OF PERIODIC VOLTAGES

4.1 INTRODUCTION

Effective control of unintentional electromagnetic radiations can be greatly facilitated when the radiations are usefully characterized. As previously noted, regulatory limits on unintentional radiations are limits on the amplitudes of their sinusoidal components, and, to verify compliance with the limits, the amplitudes of those sinusoidal components are measured. As a result, useful descriptions associate the amplitudes and frequencies of the sinusoidal components of unintentional radiations with voltage and current parameters that are either well known or easily evaluated.

In addition, electromagnetic radiations are the direct result of time-varying electric currents, and currents are sometimes difficult to observe and evaluate. However, currents are established by voltages, which are usually relatively easy to observe and evaluate with an oscilloscope. Therefore, a good start toward obtaining useful descriptions of unnecessary electromagnetic radiations is to describe time-varying voltages in terms of their sinusoidal components. Then, the currents caused by those voltages and the radiations caused by the currents can be similarly described. Given these frequency-domain descriptions, or sinusoidal component lists, the causes of the radiations can be better understood, and the radiations can be better controlled.

The objective in this chapter, then, is to provide relatively simple methods for describing commonly used periodic voltages as sums of sinusoidal voltages. These frequency-domain descriptions of the voltage waveshapes are then used to describe the currents they cause and the radiations the currents cause.

4.2 MATHEMATICAL FOUNDATIONS

The parameters that enter into the frequency-domain descriptions of periodic voltages given here will be taken from their time-domain descriptions, as observed on an oscilloscope. The parameters used in those descriptions are illustrated in Fig. 4-1. They are V_p, the peak-to-peak amplitude of the voltage; t_r, the complete (100%) rise time of the voltage; t_f, its complete fall time; t_d, the time from the start of rise to the start of fall; and T, the time of one full period. Each of these parameters will generally either be known or easily obtainable by viewing a voltage with an oscilloscope.

The mathematical bases for the development that follows are the well-known concepts of the Fourier series, which can be summarized as follows. If a periodic voltage $v(t)$ has a frequency f and a period $T = 1/f$, then it has the Fourier series representation

$$v(t) = \frac{a_0}{2} + \sum_{n=1}^{\infty} [a_n \cos(2\pi f_n t) + b_n \sin(2\pi f_n t)] \tag{4-1}$$

In this description of $v(t)$, the frequencies of the sinusoidal components are $f_n = nf = n/T$, for all integers $n \geq 0$, where, as noted above, $f = 1/T$ is the frequency of $v(t)$. And, the amplitudes of the sinusoidal components are

$$a_n = \frac{2}{T} \int_0^T v(t) \cos(2\pi f_n t)\,dt \tag{4-2}$$

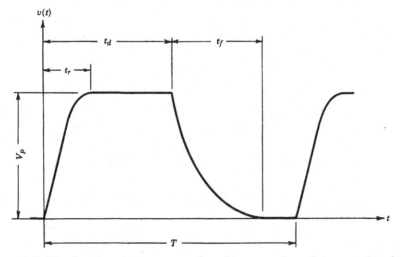

Figure 4-1. The time-domain parameters of a voltage waveshape that are used to obtain a frequency-domain description of the voltage.

and

$$b_n = \frac{2}{T} \int_0^T v(t) \sin(2\pi f_n t) dt \tag{4-3}$$

Because $f_n = nf$, for all $n \geq 1$, the frequency f is called the *fundamental frequency* of the Fourier series for $v(t)$. And a_n and b_n are called the *Fourier coefficients* of $v(t)$.

Now, when $n = 0$, then $f_n = 0$, so that $\cos(2\pi f_n t) = \cos(0) = 1$, and

$$\frac{a_0}{2} = \frac{1}{T} \int_0^T v(t) dt \equiv \overline{v(t)} \tag{4-4}$$

The integral in Eq. 4-4 is, by definition, the average value of $v(t)$. Therefore, the notation $\overline{v(t)}$ will be used here to denote that average, rather than $a_0/2$. Also, if $n = 0$, $\sin(2\pi f_n t) = \sin(0) = 0$, and, from Eq. 4-3, it follows that $b_0 = 0$. Accordingly, b_0 does not appear in Eq. 4-1.

When $n \geq 1$, it will be useful to express the summands in Eq. 4-1 somewhat differently than they are expressed there. To do that, the angle ϕ_n is defined for any a_n and b_n to be

$$\phi_n = \arctan\left(\frac{b_n}{a_n}\right) + \frac{\pi}{2}\left(1 - \frac{a_n}{|a_n|}\right) \tag{4-5}$$

As illustrated in Fig. 4-2, the second term in this definition is a necessary addition to the arctangent, to give ϕ_n a full range of 2π radians. That is, when a_n is negative, $\pi/2 \leq \phi_n \leq 3\pi/2$, and it is necessary to add π to $\arctan(b_n/a_n)$, because $-\pi/2 \leq \arctan(b_n/a_n) \leq \pi/2$.

Given this definition of ϕ_n, then, it follows that

$$\sin(\phi_n) = \frac{b_n}{\sqrt{a_n^2 + b_n^2}} \quad \text{and} \quad \cos(\phi_n) = \frac{a_n}{\sqrt{a_n^2 + b_n^2}} \tag{4-6}$$

As a result, the summands of Eq. 4-1 can be expressed as

$$a_n \cos(2\pi f_n t) + b_n \sin(2\pi f_n t)$$

$$= \sqrt{a_n^2 + b_n^2} \left[\cos(\phi_n)\cos(2\pi f_n t) + \sin(\phi_n)\sin(2\pi f_n t)\right]$$

$$= \sqrt{a_n^2 + b_n^2} \cos(2\pi f_n t - \phi_n) \tag{4-7}$$

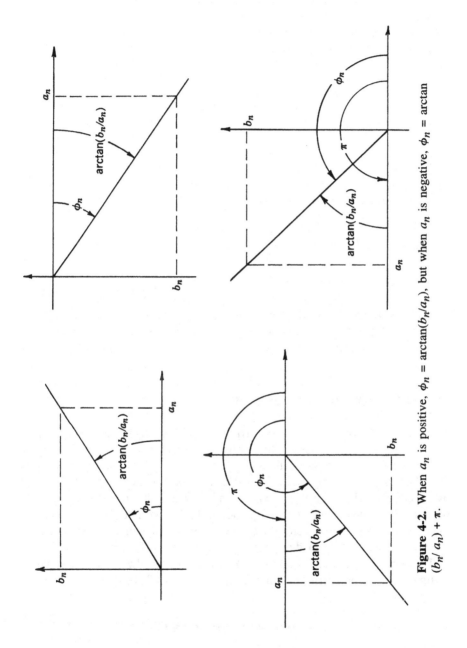

Figure 4-2. When a_n is positive, $\phi_n = \arctan(b_n/a_n)$, but when a_n is negative, $\phi_n = \arctan (b_n/a_n) + \pi$.

The latter expression in these equations follows from trigonometric identity A-3 given in Appendix A, which says that $\cos(x)\cos(y) + \sin(x)\sin(y) = \cos(x-y)$, for any x and y.

Thus, any periodic voltage $v(t)$ can be viewed as an infinite sum of sinusoidal voltages that has the general form

$$v(t) = \overline{v(t)} + \sum_{n=1}^{\infty} |V_n| \cos(2\pi f_n t - \phi_n) \tag{4-8}$$

In this equation,

$\overline{v(t)}$ = the *average value* of $v(t)$

$|V_n| = \sqrt{a_n^2 + b_n^2}$

$\phi_n = \arctan(b_n/a_n) + (\pi/2)(1 - a_n/|a_n|)$

where a_n and b_n are the Fourier coefficients of $v(t)$.

Equation 4-8 is a general description of any periodic voltage, in terms of the coefficients a_n and b_n. Therefore, attention is now turned to the evaluation of $|V_n|$ and ϕ_n for periodic voltage waveforms that are commonly used in practice. The practical advantages to be gained by doing this will soon be obvious.

4.3 BASIC VOLTAGE WAVEFORMS

A large number of commonly occurring periodic voltage waveforms can be obtained with the four basic waveforms illustrated in Fig. 4-3. A particular voltage waveform is obtained by adding the graph of the basic waveform that describes its rise to the graph of the basic waveform that describes its fall, and by specifying values for V_p, t_r, t_d, t_f, and T. The Fourier coefficients of the waveform are then found simply by adding the known Fourier coefficients of the basic waveforms that were added. Several examples will be considered after the Fourier coefficients of the basic waveforms are obtained.

The basic waveforms of Fig. 4-3 can each be described as a specific function of time, for one full period of time from $t = 0$, to $t = T$, as follows:

$$v_1(t) = \frac{V_p}{t_r} t \quad \text{for } 0 \le t \le t_r$$

$$v_1(t) = V_p \quad \text{for } t_r \le t \le t_d \tag{4-9}$$

$$v_1(t) = 0 \quad \text{for } t_d < t \le T$$

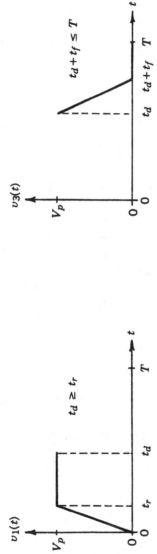

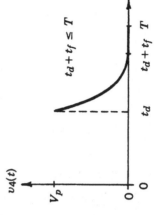

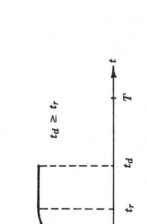

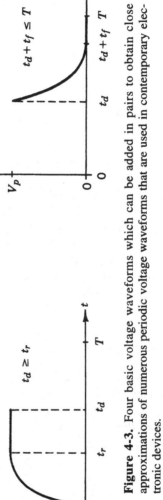

Figure 4-3. Four basic voltage waveforms which can be added in pairs to obtain close approximations of numerous periodic voltage waveforms that are used in contemporary electronic devices.

$$v_2(t) = V_p(1 - e^{-2et/t_r}) \qquad \text{for } 0 \le t \le t_r$$
$$v_2(t) = V_p \qquad\qquad\qquad \text{for } t_r \le t \le t_d \qquad\qquad (4\text{-}10)$$
$$v_2(t) = 0 \qquad\qquad\qquad \text{for } t_d < t \le T$$

$$v_3(t) = 0 \qquad\qquad\qquad\qquad \text{for } 0 \le t < t_d$$

$$v_3(t) = \frac{V_p}{t_f}(t_d + t_f - t) \qquad \text{for } t_d \le t \le t_d + t_f \qquad (4\text{-}11)$$

$$v_3(t) = 0 \qquad\qquad\qquad\qquad \text{for } t_d + t_f \le t \le T$$

$$v_4(t) = 0 \qquad\qquad\qquad\qquad \text{for } 0 \le t < t_d$$
$$v_4(t) = V_p e^{-2e(t - t_d)/t_f} \qquad \text{for } t_d \le t \le t_d + t_f \qquad (4\text{-}12)$$
$$v_4(t) = 0 \qquad\qquad\qquad\qquad \text{for } t_d + t_f \le t \le T$$

In all of the above expressions, and in subsequent expressions in which it appears, $e = 2.71828\ldots$, the base of natural logarithms.

With these *time-domain* expressions for the basic voltage waveforms from $t = 0$ to $t = T$, their Fourier coefficients can now be obtained by integration, as indicated in Eqs. 4-2 and 4-3. Details of those integrations are summarized in Appendix B. The results are given below with additional subscripts of 1, 2, 3, and 4 added to a_n and b_n to identify which basic voltage waveform they represent. To simplify these expressions and other expressions to come the following parameters are defined:

$$t_R \equiv \frac{t_r}{e} \quad \text{and} \quad t_F \equiv \frac{t_f}{e}$$

Note also that $\omega_n = 2\pi f_n = 2\pi n f = 2\pi n/T$. These expressions will all be used interchangeably.

The Fourier coefficients for the basic voltage waveform $v_1(t)$, which describes *linear rises*, are

$$a_{n1} = -\frac{V_p T}{n^2 \pi^2} \frac{\sin^2(\pi f_n t_r)}{t_r} + \frac{V_p}{n\pi} \sin(\omega_n t_d)$$

and

$$b_{n1} = \frac{V_p T}{n^2 \pi^2} \frac{\sin(\pi f_n t_r)\cos(\pi f_n t_r)}{t_r} - \frac{V_p}{n\pi} \cos(\omega_n t_d) \qquad (4\text{-}13)$$

The Fourier coefficients for the basic voltage waveform $v_2(t)$, which describes *exponential rises*, are

$$a_{n2} = -\frac{V_p T}{n^2 \pi^2} \frac{\sin^2(\alpha_n)}{t_R} + \frac{V_p}{n\pi} \sin(\omega_n t_d)$$

and

$$b_{n2} = \frac{V_p T}{n^2 \pi^2} \frac{\sin(\alpha_n) \cos(\alpha_n)}{t_R} - \frac{V_p}{n\pi} \cos(\omega_n t_d) \qquad (4\text{-}14)$$

where

$$\alpha_n = \arctan(\pi f_n t_R) A$$

$$\sin(\alpha_n) = \frac{\pi f_n t_R}{\sqrt{1 + (\pi f_n t_R)^2}}$$

and

$$\cos(\alpha_n) = \frac{1}{\sqrt{1 + (\pi f_n t_R)^2}} \qquad (4\text{-}15)$$

The Fourier coefficients for the basic voltage waveform $v_3(t)$, which describes linear falls, are

$$a_{n3} = \frac{V_p T}{n^2 \pi^2} \frac{\sin(\pi f_n t_f) \sin(\omega_n t_d + \pi f_n t_f)}{t_f} - \frac{V_p}{n\pi} \sin(\omega_n t_d)$$

and

$$b_{n3} = -\frac{V_p T}{n^2 \pi^2} \frac{\sin(\pi f_n t_f) \cos(\omega_n t_d + \pi f_n t_f)}{t_f} + \frac{V_p}{n\pi} \cos(\omega_n t_d) \qquad (4\text{-}16)$$

And, the Fourier coefficients for the basic voltage waveform $v_4(t)$, which describes exponential falls, are

$$a_{n4} = \frac{V_p T}{n^2 \pi^2} \frac{\sin(\beta_n) \sin(\omega_n t_d + \beta_n)}{t_F} - \frac{V_p}{n\pi} \sin(\omega_n t_d)$$

and

$$b_{n4} = -\frac{V_p T}{n^2 \pi^2} \frac{\sin(\beta_n) \cos(\omega_n t_d + \beta_n)}{t_F} + \frac{V_p}{n\pi} \cos(\omega_n t_d) \qquad (4\text{-}17)$$

where

$$\beta_n = \arctan(\pi f_n t_F)$$

$$\sin(\beta_n) = \frac{\pi f_n t_F}{\sqrt{1 + (\pi f_n t_F)^2}}$$

and

$$\cos(\beta_n) = \frac{1}{\sqrt{1 + (\pi f_n t_F)^2}} \tag{4-18}$$

Note, for future reference, that the angle α_n is associated with exponential rises, and the angle β_n is associated with exponential falls.

Before applying the above results, it is worthwhile to observe some similarities in the expressions obtained. For example, $v_1(t)$ and $v_2(t)$ are both basic waveforms that describe rises. And their Fourier coefficients a_{n1}, which is associated with linear rises, and a_{n2}, which is associated with exponential rises, are the same except that in a_{n2}, t_R replaces t_r, and $\alpha_n = \arctan(\pi f_n t_R)$ replaces $\pi f_n t_r$. The same is also true for the coefficients b_{n1} and b_{n2}.

Similarly, $v_3(t)$ and $v_4(t)$ are both basic waveforms that describe falls. Their Fourier coefficients a_{n3}, which is associated with linear falls, and a_{n4}, which is associated with exponential falls, are the same except that in a_{n4}, t_F replaces t_f, and $\beta_n = \arctan(\pi f_n t_F)$ replaces $\pi f_n t_f$. The same is also true for the coefficients b_{n3} and b_{n4}.

Finally, the average values of the basic voltage waveforms for rises are

$$\overline{v_1(t)} = \frac{V_p}{2T} (2t_d - t_r) \quad \text{and} \quad \overline{v_2(t)} = \frac{V_p}{2T} (2t_d - t_R)$$

And, the average values of the basic voltage waveforms for falls are

$$\overline{v_3(t)} = \frac{V_p}{2T} t_f \quad \text{and} \quad \overline{v_4(t)} = \frac{V_p}{2T} t_F \tag{4-19}$$

These values are of relatively little importance here, because average values of voltages do not cause time-varying currents. However, they are essential to fully convert a given time-domain description of a voltage into its equal frequency-domain description.

4.4 COMMONLY OCCURRING PERIODIC VOLTAGE WAVEFORMS

Any periodic voltage can be described as a sum of sinusoidal voltages using the Fourier coefficients a_n and b_n. And a large number of periodic voltages that are commonly used in contemporary electrical engineering practice can be described in the time domain as sums of either $v_1(t)$ or $v_2(t)$ added to either $v_3(t)$ or $v_4(t)$. As a result of their similar construction, these voltages also have similar frequency-domain descriptions.

These voltage waveshapes can be divided into four general categories: (1) those that have a linear rise and fall, (2) those that have an exponential rise and fall, (3) those that have a linear rise and an exponential fall, and (4) those that have an exponential rise and a linear fall. Each of these categories is also divisible into four subcategories as a result of the relationships $t_r \le t_d$ and $t_d + t_f \le T$. In other words, either $t_r = t_d$ or $t_r < t_d$ and either $t_d + t_f < T$ or $t_d + t_f = T$. Thus, there are four different subcategories of the four categories, making a total of 16 basically different kinds of periodic voltage waveshapes to be considered here.

Each of these voltage waveshape categories is now described and examined in detail, and several examples are given. Readers are encouraged to plot the examples on a calculator or computer to verify the results for themselves.

4.5 VOLTAGES WITH A LINEAR RISE AND A LINEAR FALL

Periodic voltage waveforms that have both a linear rise and a linear fall can have any one of the four general waveforms illustrated in Fig. 4-4. The waveforms are different, because of the relationships $t_r \le t_d$ and $t_d + t_f \le T$. However, each of the different voltage waveshapes has the same general description as a sum of sinusoidal voltages in terms of the variables t_r, t_d, t_f, T, and V_p. To see this, notice that each of the waveshapes in Fig. 4-4 can be described as $v(t) = v_1(t) + v_3(t)$. This can be done because $v_1(t)$ is the basic waveshape for linear rises and $v_3(t)$ is the basic waveshape for linear falls. And because $v(t) = v_1(t) + v_3(t)$, the Fourier coefficients of $v(t)$ are $a_n = a_{n1} + a_{n3}$ and $b_n = b_{n1} + b_{n3}$. Therefore, it follows from Eqs. 4-13 and 4-16 that

$$a_n = -\frac{V_p T}{n^2 \pi^2} \left[\frac{\sin^2(\pi f_n t_r)}{t_r} - \frac{\sin(\pi f_n t_f) \sin(\omega_n t_d + \pi f_n t_f)}{t_f} \right]$$

and

$$b_n = \frac{V_p T}{n^2 \pi^2} \left[\frac{\sin(\pi f_n t_r) \cos(\pi f_n t_r)}{t_r} - \frac{\sin(\pi f_n t_f) \cos(\omega_n t_d + \pi f_n t_f)}{t_f} \right] \quad (4\text{-}20)$$

Based on these equations, since $|V_n| = \sqrt{a_n^2 + b_n^2}$, it also follows that

$$|V_n| = \frac{V_p T}{n^2 \pi^2} D_n \quad (4\text{-}21)$$

where

$$D_n = \sqrt{ \frac{\sin^2(\pi f_n t_r)}{t_r^2} - 2 \frac{\sin(\pi f_n t_r) \sin(\pi f_n t_f)}{t_r t_f} \cos[\pi f_n (2 t_d - t_r + t_f)] + \frac{\sin^2(\pi f_n t_f)}{t_f^2} }$$

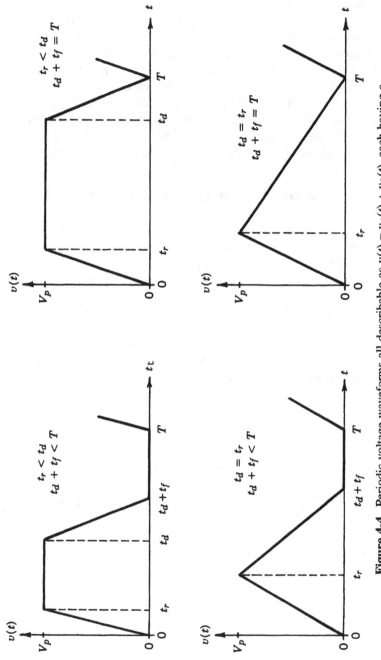

Figure 4-4. Periodic voltage waveforms all describable as $v(t) = v_1(t) + v_3(t)$, each having a linear rise and a linear fall.

This last quantity, D_n, is called the *transition factor*. The development of this expression for D_n from Eq. 4-20 is detailed in Section A.3 of Appendix A. In that development, for this class of voltages, $t_1 = t_r$, $t_2 = t_f$, $x = \pi f_n t_r$, $y = \pi f_n t_f$, and $z = \pi f_n(2t_d + t_f)$.

In other words, from the above and from Eq. 4-8, it is seen that all of the voltage waveshapes in Fig. 4-4 have the general frequency-domain representation

$$v(t) = \overline{v(t)} + \sum_{n=1}^{\infty} \frac{V_p T}{n^2 \pi^2} D_n \cos(2\pi f_n t - \phi_n) \tag{4-22}$$

where $D_n = D_n(t_r, t_f, t_d, f_n)$ is given in Eqs. 4-21.

The general expression for the average value of these voltages, from Eqs. 4-19, is

$$\overline{v(t)} = \overline{v_1(t)} + \overline{v_3(t)} = \frac{V_p}{2T} (2t_d - t_r + t_f) \tag{4-23}$$

Two specific examples from this voltage category are now examined.

Example 4-1   Suppose the values of $|V_n|$ and ϕ_n are wanted for the trapezoidal waveshape of Fig. 4-5. In that figure, $t_r = T/10$, $t_d = T/2$, $t_f = T/4$, and $V_p = 10$. (Precise units of time and voltage are of no immediate concern.) Substituting these values in Eqs. 4-20 yields

$$a_n = -\frac{10}{n^2 \pi^2} \left[10 \sin^2 \left(\frac{n\pi}{10} \right) - 4 \sin \left(\frac{n\pi}{4} \right) \sin \left(\frac{5n\pi}{4} \right) \right]$$

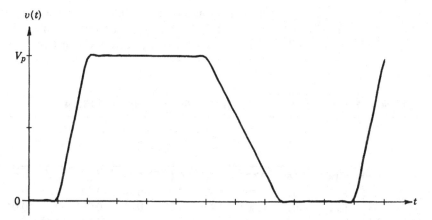

Figure 4-5. The periodic voltage waveshape of Example 4-1 for which $t_r = T/10$, $t_d = T/2$, and $t_f = T/4$.

and

$$b_n = \frac{10}{n^2\pi^2}\left[10\sin\left(\frac{n\pi}{10}\right)\cos\left(\frac{n\pi}{10}\right) - 4\sin\left(\frac{n\pi}{4}\right)\cos\left(\frac{5n\pi}{4}\right)\right] \quad (4\text{-}24)$$

Therefore,

$$\phi_n = \arctan\left(\frac{b_n}{a_n}\right) + \frac{\pi(1 - a_n/|a_n|)}{2}$$

can be determined. And $|V_n|$ can be evaluated either with $|V_n| = \sqrt{a_n^2 + b_n^2}$, or with the equation

$$|V_n| = \frac{10}{n^2\pi^2} D_n \quad (4\text{-}25)$$

where

$$D_n = 4\sqrt{6.25\sin^2\left(\frac{n\pi}{10}\right) - 5\sin\left(\frac{n\pi}{10}\right)\sin\left(\frac{n\pi}{4}\right)\cos(1.15n\pi) + \sin^2\left(\frac{n\pi}{4}\right)}.$$

Thus are the values of $|V_n|$ and ϕ_n obtained for this voltage waveshape. Their first 10 values are given in Table 4-1, and a plot of the voltage in the frequency domain is given in Fig. 4-6, for the first 10 values of $|V_n|$ and f_n. Voltage plots in the frequency domain are sometimes useful to determine how rapidly $|V_n|$ decreases with frequency.

The average value of the voltage is $\overline{v(t)} = (V_p/2T)(wt_d - t_r + t_f) = 5.75$.

Example 4.2 Suppose values of $|V_n|$ and ϕ_n are wanted for the triangular voltage waveshape illustrated in Fig. 4-7. As indicated in the figure, $V_p = 10$, $t_r = t_d = T/5$, and $t_f = 4T/5$. When these values are substituted in Eqs. 4-20, it is seen that

TABLE 4-1. Frequency Domain Parameters for the Voltage of Fig. 4-5

	\multicolumn{10}{c}{n}
	1	2	3	4	5	6	7	8	9	10
a_n	−2.99	0.14	−0.96	−0.57	−0.49	−0.14	−0.18	−0.05	−0.04	0.04
b_n	5.00	1.20	0.31	0.19	0.08	−0.08	−0.14	−0.08	−0.01	0
$|V_n|$	5.83	1.21	1.01	0.60	0.49	0.16	0.23	0.09	0.04	0.04
ϕ_n	2.11	1.46	2.83	2.83	2.98	3.67	3.81	4.08	3.45	0

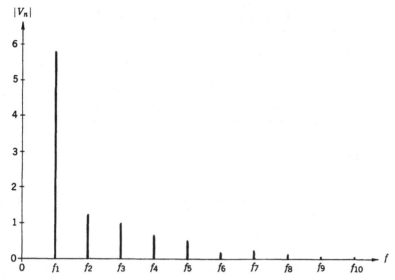

Figure 4-6. A frequency-domain representation of the trapezoidal voltage waveshape of Example 4-1 and Fig. 4-5.

$$a_n = -\frac{10}{n^2\pi^2} \left[5 \sin^2\left(\frac{n\pi}{5}\right) - 1.25 \sin\left(\frac{4n\pi}{5}\right) \sin\left(\frac{6n\pi}{5}\right) \right]$$

and

$$b_n = \frac{10}{n^2\pi^2} \left[5 \sin\left(\frac{n\pi}{5}\right) \cos\left(\frac{n\pi}{5}\right) - 1.25 \sin\left(\frac{4n\pi}{5}\right) \cos\left(\frac{6n\pi}{5}\right) \right] \quad (4\text{-}26)$$

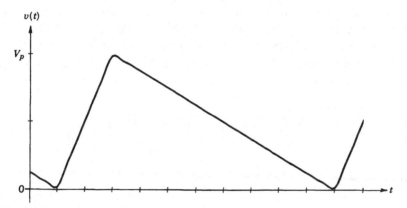

Figure 4-7. The periodic voltage waveshape of Example 4-2 for which $t_r = t_d = T/5$, and $t_f = 4T/5$.

TABLE 4-2. Values of $|V_n|$ and ϕ_n for the Voltage of Fig. 4-7

	n									
	1	2	3	4	5	6	7	8	9	10
$|V_n|$	3.72	1.51	0.67	0.23	0	0.10	0.12	0.09	0.05	0
ϕ_n	2.20	2.83	3.46	4.08	—	2.20	2.83	3.46	4.08	—

And from Eqs. 4-21,

$$|V_n| = \frac{10}{n^2\pi^2} D_n \tag{4-27}$$

where

$$D_n = 5\sqrt{\sin^2\left(\frac{n\pi}{5}\right) - 0.5\sin\left(\frac{n\pi}{5}\right)\sin\left(\frac{4n\pi}{5}\right)\cos(n\pi) + 0.0625\sin^2\left(\frac{4n\pi}{5}\right)}$$

Given these expressions, the values of $|V_n|$ and ϕ_n can again be readily determined. The first 10 values of each are given in Table 4-2 and for plotting purposes, $\overline{v(t)} = (V_p/2T)(2t_d - t_r + t_f) = 5$.

4.6 VOLTAGES WITH AN EXPONENTIAL RISE AND AN EXPONENTIAL FALL

Periodic voltage waveforms with both an exponential rise and an exponential fall can have any one of the four general forms illustrated in Fig. 4-8. Using the basic waveforms given in Fig. 4-3, any of these voltages can be described as $v(t) = v_2(t) + v_4(t)$. Thus, any one of these voltages has Fourier coefficients of the general form $a_n = a_{n2} + a_{n4}$ and $b_n = b_{n2} + b_{n4}$. Referring back to Eqs. 4-14 and 4-17, it can be seen that

$$a_n = -\frac{V_pT}{n^2\pi^2}\left[\frac{\sin^2(\alpha_n)}{t_R} - \frac{\sin(\beta_n)\sin(\omega_n t_d + \beta_n)}{t_F}\right]$$

and

$$b_n = \frac{V_pT}{n^2\pi^2}\left[\frac{\sin(\alpha_n)\cos(\alpha_n)}{t_R} - \frac{\sin(\beta_n)\cos(\omega_n t_d + \beta_n)}{t_F}\right] \tag{4-28}$$

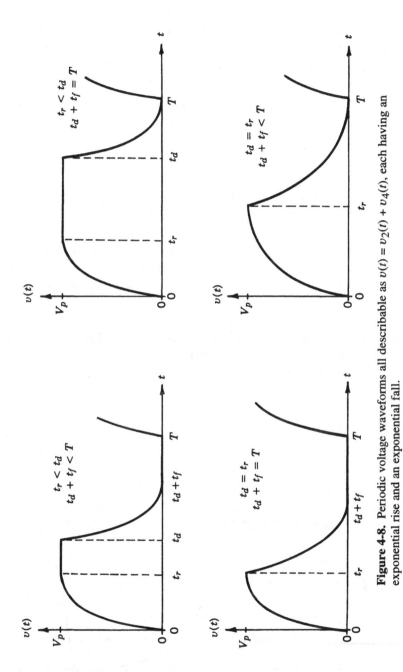

Figure 4-8. Periodic voltage waveforms all describable as $v(t) = v_2(t) + v_4(t)$, each having an exponential rise and an exponential fall.

Notice the similarity in form between these equations and Eqs. 4-20, which describe voltages with a linear rise and fall. The only differences are that, in Eqs. 4-28, t_R and t_F have replaced t_r and t_f, and $\alpha_n = \arctan(\pi f_n t_R)$ and $\beta_n = \arctan(\pi f_n t_F)$ have replaced $\pi f_n t_r$ and $\pi f_n t_f$.

Since $|V_n| = \sqrt{a_n^2 + b_n^2}$, based on Eqs. 4-28 it follows that

$$|V_n| = \frac{V_p T}{n^2 \pi^2} D_n \qquad (4\text{-}29)$$

where

$$D_n = \sqrt{\frac{\sin^2(\alpha_n)}{t_R^2} - 2\frac{\sin(\alpha_n)\sin(\beta_n)}{t_R t_F}\cos(\omega_n t_d - \alpha_n + \beta_n) + \frac{\sin^2(\beta_n)}{t_F^2}}$$

Details of the derivation of this expression for the transition factor D_n are given in Appendix A. In this case, the variable assignment is $t_1 = t_R$, $t_2 = t_F$, $x = \alpha_n$, $y = \beta_n$, and $z = \omega_n t_d + \beta_n$. These voltages have an average value of $v(t) = (V_p/2T)(2t_d - t_R + t_F)$.

Example 4-3 Consider the periodic voltage illustrated in Fig. 4-9. The voltage has an exponential rise and an exponential fall, and $t_r = T/4$, $t_d = t_f = T/2$, and $V_p = 10$. Therefore, $t_R = T/4e$ and $t_F = T/2e$, and from Eq. 4-28 it follows that

$$a_n = -\frac{20e}{n^2\pi^2}[2\sin^2(\alpha_n) - \sin(\beta_n)\sin(n\pi + \beta_n)]$$

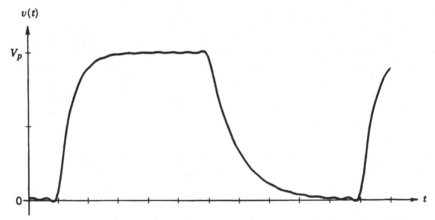

Figure 4-9. The periodic voltage waveshape of Example 4-3 for which $t_R = T/4e$, $t_F = T/2e$, and $t_d = T/2$.

TABLE 4-3. Values of $|V_n|$ and ϕ_n for the Voltage of Fig. 4-9

					n					
	1	2	3	4	5	6	7	8	9	10
$|V_n|$	5.77	0.52	1.31	0.24	0.57	0.13	0.31	0.08	0.19	0.05
ϕ_n	1.97	1.38	2.42	2.02	2.64	2.34	2.76	2.52	2.84	2.64

and

$$b_n = \frac{20e}{n^2\pi^2}\,[2\sin(\alpha_n)\cos(\alpha_n) - \sin(\beta_n)\cos(n\pi + \beta_n)]$$

Also,

$$|V_n| = \frac{10}{n^2\pi^2}\,D_n \tag{4-30}$$

where

$$D_n = 2e\sqrt{4\sin^2(\alpha_n) - 4\sin(\alpha_n)\sin(\beta_n)\cos(n\pi - \alpha_n + \beta_n) + \sin^2(\beta_n)}$$

and $\alpha_n = \arctan(n\pi/4e)$, and $\beta_n = \arctan(n\pi/2e)$.

$|V_n|$ and ϕ_n can now be found for the voltage in Fig. 4-9 using these equations. The values of those parameters for the first 10 values of n are given below in Table 4-3. Also, the average value of the voltage is $\overline{v(t)} = 5(1 - 1/4e + 1/2e)$ = 5.46.

4.7 VOLTAGES WITH A LINEAR RISE AND AN EXPONENTIAL FALL

Periodic voltages that rise linearly and fall exponentially will have one of the four general time-domain waveforms shown in Fig. 4-10. Using the basic waveforms, each of these voltages can be described as $v(t) = v_1(t) + v_4(t)$, with the Fourier coefficients $a_n = a_{n1} + a_{n4}$ and $b_n = b_{n1} + b_{n4}$. Therefore,

$$a_n = -\frac{V_pT}{n^2\pi^2}\left[\frac{\sin^2(\pi f_n t_f)}{t_r} - \frac{\sin(\beta_n)\sin(\omega_n t_d + \beta_n)}{t_F}\right]$$

$$b_n = \frac{V_pT}{n^2\pi^2}\left[\frac{\sin(\pi f_n t_r)\cos(\pi f_n t_r)}{t_r} - \frac{\sin(\beta_n)\cos(\omega_n t_d + \beta_n)}{t_F}\right]$$

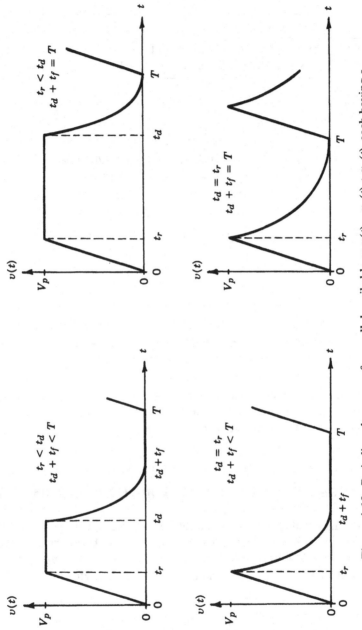

Figure 4-10. Periodic voltage waveforms all describable as $v(t) = v_1(t) + v_4(t)$, each having a linear rise and an exponential fall.

and

$$|V_p| = \frac{V_p T}{n^2 \pi^2} D_n \qquad (4\text{-}31)$$

where

$$D_n = \sqrt{\frac{\sin^2(\pi f_n t_r)}{t_r^2} - 2\frac{\sin(\pi f_n t_r)\sin(\beta_n)}{t_r t_F}\cos(\omega_n t_d - \pi f_n t_r + \beta_n) + \frac{\sin^2(\beta_n)}{t_F^2}}$$

And, from Eqs. 4-19, the average value of these voltages is $v(t) = (V_p/2T)(2t_d - t_r + t_F)$.

Example 4-4 Consider the voltage waveshape shown in Fig. 4-11. It has a linear rise and an exponential fall, $t_r = t_d = T/10$, $t_f = 9T/10$, and $V_p = 10$. Therefore, $t_F = 9T/10e$, and the Fourier coefficients of the voltage are

$$a_n = -\frac{100}{n^2 \pi^2}\left[\sin^2\left(\frac{n\pi}{10}\right) - \frac{e}{9}\sin(\beta_n)\sin\left(\frac{n\pi}{5} + \beta_n\right)\right]$$

and

$$b_n = \frac{100}{n^2 \pi^2}\left[\sin\left(\frac{n\pi}{10}\right)\cos\left(\frac{n\pi}{10}\right) - \frac{e}{9}\sin(\beta_n)\cos\left(\frac{n\pi}{5} + \beta_n\right)\right] \qquad (4\text{-}32)$$

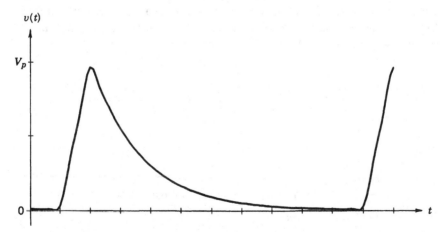

Figure 4-11. The periodic voltage waveshape of Example 4-4 for which $t_r = t_d = T/10$, and $t_F = 9T/10e$.

TABLE 4-4. Values of $|V_n|$ and ϕ_n for the Voltage of Fig. 4-11

	1	2	3	4	5	6	7	8	9	10
					n					
$|V_n|$	2.94	1.75	1.13	0.77	0.52	0.35	0.22	0.13	0.06	0.03
ϕ_n	1.14	1.80	2.28	2.70	3.10	3.49	3.90	4.35	-1.32	-0.10

Therefore, the transition factor is

$$D_n = 10\sqrt{\sin^2\left(\frac{n\pi}{10}\right) - 2e\sin\left(\frac{n\pi}{10}\right)\sin(\beta_n)\cos\left(\frac{n\pi}{5} + \beta_n\right) + \frac{e^2}{81}\sin^2(\beta_n)}$$

and

$$|V_n| = \frac{10}{n^2\pi^2} D_n \tag{4-33}$$

Based on these equations, the first 10 values of $|V_n|$ and ϕ_n are those given in Table 4-4. For plotting purposes, $\overline{v(t)} = 2.16$.

4.8 VOLTAGES WITH AN EXPONENTIAL RISE AND A LINEAR FALL

Periodic voltages that rise exponentially and fall linearly can have any one of the four general forms illustrated in Fig. 4-12. Referring once again to the basic waveforms of Fig. 4-3 and their equations, it is seen that the four voltages of Fig. 4-12 can be described as $v(t) = v_2(t) + v_3(t)$. Therefore, $a_n = a_{n2} + a_{n3}$ and $b_n = b_{n2} + b_{n3}$, and Eqs. 4-14 and 4-16 give

$$a_n = -\frac{V_pT}{n^2\pi^2}\left[\frac{\sin^2(\alpha_n)}{t_R} - \frac{\sin(\pi f_n t_f)\sin(\omega_n t_d + \pi f_n t_f)}{t_f}\right]$$

and

$$b_n = \frac{V_pT}{n^2\pi^2}\left[\frac{\sin(\alpha_n)\cos(\alpha_n)}{t_R} - \frac{\sin(\pi f_n t_f)\cos(\omega_n t_d + \pi f_n t_f)}{t_f}\right] \tag{4-34}$$

Therefore,

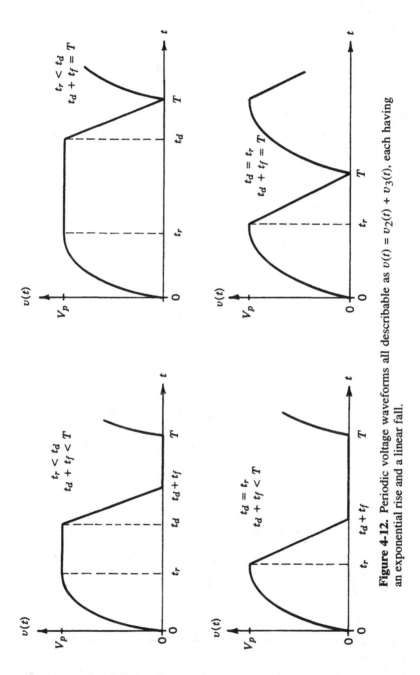

Figure 4-12. Periodic voltage waveforms all describable as $v(t) = v_2(t) + v_3(t)$, each having an exponential rise and a linear fall.

$$D_n = \sqrt{\frac{\sin^2(\alpha_n)}{t_R^2} - 2\frac{\sin(\alpha_n)\sin(\pi f_n t_f)}{t_R t_f} \cos(\omega_n t_d - \alpha_n + \pi f_n t_f) + \frac{\sin^2(\pi f_n t_f)}{t_f^2}}$$

In all of these equations, $\alpha_n = \arctan(\pi f_n t_R)$ and the average value of $v(t)$ is $\overline{v(t)} = (V_p/2T)(2t_d - t_R + t_f)$.

Example 4-5 Consider the waveshape of Fig. 4-13. Given that $t_r = T/4$, $t_d = T/2$, $t_f = T/8$, and $V_p = 10$, it follows that $t_R = T/4e$ and $\alpha_n = \arctan(n\pi/4e)$ from Eqs. 4-28, and that

$$a_n = -\frac{40}{n^2\pi^2}\left[e\sin^2(\alpha_n) - 2\sin\left(\frac{n\pi}{8}\right)\sin\left(\frac{9n\pi}{8}\right)\right]$$

$$b_n = \frac{40}{n^2\pi^2}\left[e\sin(\alpha_n)\cos(\alpha_n) - 2\sin\left(\frac{n\pi}{8}\right)\cos\left(\frac{9n\pi}{8}\right)\right] \quad (4\text{-}35)$$

Also,

$$|V_n| = \frac{10}{n^2\pi^2} D_n \quad (4\text{-}36)$$

where

$$D_n = 4\sqrt{e\sin^2(\alpha_n) - 4\sin(\alpha_n)\sin\left(\frac{n\pi}{8}\right)\cos\left(\frac{9n\pi}{8} - \alpha_n\right) + 4\sin^2\left(\frac{n\pi}{8}\right)}$$

The first 10 values of $|V_n|$ and ϕ_n are given in Table 4-5, and $\overline{v(t)} = 5.17$.

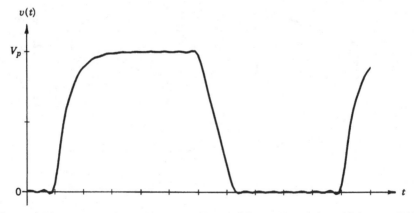

Figure 4-13. The periodic voltage waveshape of Example 4-5 for which $t_R = T/4e$, $t_f = T/8$, and $t_d = T/2$.

TABLE 4-5. Values of $|V_n|$ and ϕ_n for the Voltage of Fig. 4-13

					n					
	1	2	3	4	5	6	7	8	9	10
$|V_p|$	6.15	0.37	1.59	0.36	0.58	0.27	0.21	0.16	0.16	0.06
ϕ_n	1.91	0.51	2.52	1.25	2.98	2.02	2.99	2.73	2.73	3.25

4.9 CONSOLIDATION

All of the voltage waveforms considered above have the general frequency domain description

$$v(t) = \overline{v(t)} + \sum_{n=1}^{\infty} |V_n| \cos(2\pi f_n t - \phi_n)$$

in which

$$|V_n| = \sqrt{a_n^2 + b_n^2}$$

and

$$\phi_n = \arctan\left(\frac{b_n}{a_n}\right) + \frac{\pi}{2}\left(1 - \frac{a_n}{|a_n|}\right) \tag{4-37}$$

where a_n and b_n are the the Fourier coefficients of $v(t)$. For each of the 16 categories of voltage waveforms considered, the Fourier coefficients a_n are all quite similar to one another, as are the coefficients b_n. In fact, with only a few additional generalizations, the similarities can be extended to yield one expression for a_n and one expression for b_n for all of those voltage waveforms. This is done as follows.

First, if the rise is linear, let $t_R = t_r$ and let $A_n = \pi f_n t_r$. If the rise is exponential, let $t_R = t_r/e$ and let $A_n = \alpha_n = \arctan(\pi f_n t_R)$. If the fall is linear, let $t_F = t_f$ and let $B_n = \pi f_n t_f$. If the fall is exponential, let $t_F = t_f/e$ and let $B_n = \beta_n = \arctan(\pi f_n t_F)$. Based on these definitions, the Fourier coefficients of any of the periodic voltage waveforms discussed above can be written as

$$a_n = -\frac{V_p T}{n^2 \pi^2}\left[\frac{\sin^2(A_n)}{t_R} - \frac{\sin(B_n)\sin(\omega_n t_d + B_n)}{t_F}\right]$$

and

$$b_n = \frac{V_p T}{n^2 \pi^2} \left[\frac{\sin(A_n)\cos(A_n)}{t_R} - \frac{\sin(B_n)\cos(\omega_n t_d + B_n)}{t_F} \right] \qquad (4\text{-}38)$$

From these expressions, it then follows that

$$|V_n| = \sqrt{a_n^2 + b_n^2}$$

$$= \frac{V_p T}{n^2 \pi^2} \sqrt{\frac{\sin^2(A_n)}{t_R^2} - 2\,\frac{\sin(A_n)}{t_R}\,\frac{\sin(B_n)}{t_F}\,\cos(\omega_n t_d - A_n + B_n) + \frac{\sin^2(B_n)}{t_F^2}}$$

$$(4\text{-}39)$$

Thus, the transition factor has the general form

$$D_n = \sqrt{\frac{\sin^2(A_n)}{t_R^2} - 2\,\frac{\sin(A_n)}{t_R}\,\frac{\sin(B_n)}{t_F}\,\cos(\omega_n t_d - A_n + B_n) + \frac{\sin^2(B_n)}{t_F^2}}$$

$$(4\text{-}40)$$

and $|V_n|$ can be more concisely described as

$$|V_n| = \frac{V_p T D_n}{n^2 \pi^2} \qquad (4\text{-}41)$$

Similarly, it follows from Eqs. 4-37 and 4-38 that

$$\phi_n = \arctan \left[\frac{t_F \sin(A_n)\cos(A_n) - t_R \sin(B_n)\cos(\omega_n t_d + B_n)}{t_F \sin^2(A_n) - t_R \sin(B_n)\sin(\omega_n t_d + B_n)} \right]$$

$$+ \frac{\pi}{2} \left[1 + \frac{t_F \sin^2(A_n) - t_R \sin(B_n)\sin(\omega_n t_d + B_n)}{|t_F \sin^2(A_n) - t_R \sin(B_n)\sin(\omega_n t_d + B_n)|} \right] \qquad (4\text{-}42)$$

Equations 4-39 and 4-42 are general expressions for the amplitudes and the phases of the frequency-domain components of all of the periodic voltages considered in this chapter. The practical significance of these expressions is that they are functions of only time-domain variables and n. In other words, $|V_n|$ and ϕ_n can be found without first having to find the Fourier coefficients, a_n and b_n. The values of V_p, t_r, t_f, t_d, and T can be found, for example, by simply viewing $v(t)$ on an oscilloscope and measuring them. Then $|V_n|$ and ϕ_n can be found for any value of n by calculating t_R, t_F, A_n, B_n, and D_n and substituting those values, together with the values of V_p, T, and t_d, in Eq. 4-38 or in Eqs. 4-39 and 4-40 and in Eq. 4-42.

4.10 SUMMARY

The purpose of this chapter is to provide a simple, practical method for obtaining frequency-domain descriptions of time-domain voltage waveforms. Sixteen different categories of periodic, time-varying voltages $v(t)$, examples of each of which are shown in Fig. 4-14, were considered here. It was seen that any of those voltages is easily describable as a sum of sinusoidal voltages of different frequencies using only parameters obtained from their time-domain descriptions. The procedure is probably best summarized with a simple example.

Suppose the amplitudes $|V_n|$ of some of the sinusoidal components of the voltage waveform $v(t)$ of Fig. 4-15a are needed. They can be found as follows. The period of that waveform is $T = 1/f$, where f is its frequency. From Fig. 4-15a it is seen that the amplitude of the voltage is $V_p = 10$ volts, its total linear rise time is $t_r = T/4$, its total exponential fall time is $t_f = T/2$, and the time from the start of t_r to the start of t_f is $t_d = T/2$.

Now, the expression developed in this chapter with which $|V_n|$ can be evaluated for any given n is

$$|V_n| = \frac{V_p T}{n^2 \pi^2} \sqrt{\frac{\sin^2(A_n)}{t_R^2} - 2\frac{\sin(A_n)}{t_R}\frac{\sin(B_n)}{t_F}\cos(2\pi n f t_d - A_n + B_n) + \frac{\sin^2(B_n)}{t_F^2}}$$

In this expression, because the rise is linear, $t_R = t_f$ and $A_n = n\pi f t_R$. Also, because the fall is exponential, $t_F = t_f/e$ and $B_n = \arctan(n\pi f t_F)$, where $e = 2.718...$. Therefore, for this voltage waveform,

$$t_R = \frac{T}{4} \qquad A_n = \frac{n\pi}{4} \qquad t_F = \frac{T}{2e} \qquad B_n = \arctan\left(\frac{n\pi}{2e}\right)$$

and

$$\cos(2\pi n f t_d - A_n + B_n) = \cos\left[\frac{3n\pi}{4} + \arctan\left(\frac{n\pi}{2e}\right)\right]$$

Substituting these values in the above expression for $|V_n|$ and solving the resulting equation for $n = 1$ to $n = 10$ yields the following table of values:

					n					
	1	2	3	4	5	6	7	8	9	10
$|V_n|$	5.57	0.72	0.69	0.32	0.32	0.05	0.15	0.08	0.10	0.02

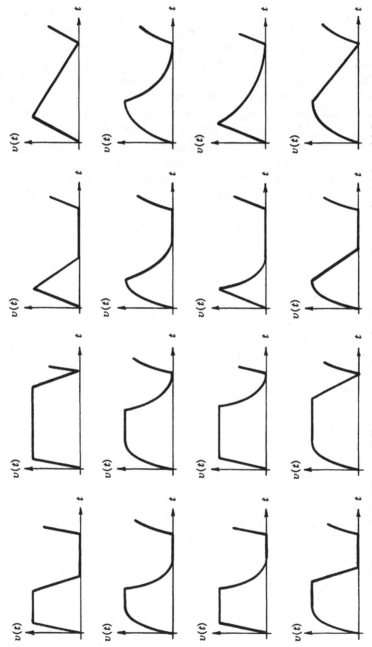

Figure 4-14. Sixteen different categories of voltage waveshapes whose sinusoidal components have easily determined amplitudes.

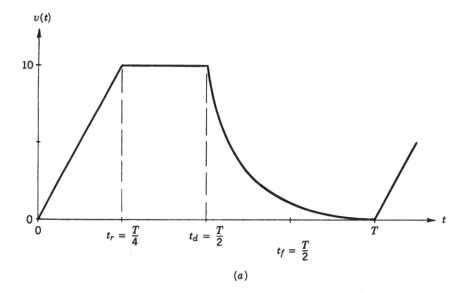

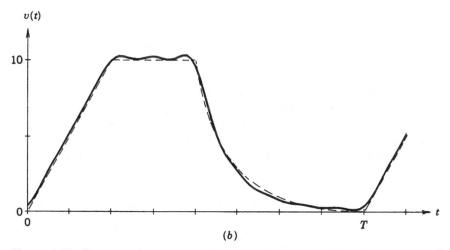

Figure 4-15. Time-domain representations of (*a*) the voltage $v(t)$ and (*b*) the sum of the first 10 frequency-domain components (—) of $v(t)$ (---).

In this way, then, the value of $|V_n|$ can be found for any frequency $f_n = nf$.

Typically, for purposes of predicting radiations, only the amplitudes of sinusoidal components will be of interest. However, to verify that the values obtained for the amplitudes $|V_n|$ are valid, the frequency-domain description can be plotted in the time domain and compared to the given voltage waveform. To do that the values of ϕ_n, the phase angles of the sinusoidal components, must be known. The expression developed here with which ϕ_n can be evaluated is

$$\phi_n = \arctan \left[\frac{t_F \sin(A_n)\cos(A_n) - t_R \sin(B_n)\cos(2\pi n f t_d + B_n)}{t_F \sin^2(A_n) - t_R \sin(B_n)\sin(2\pi n f t_d + B_n)} \right]$$

$$+ \frac{\pi}{2} \left[1 + \frac{t_F \sin^2(A_n) - t_R \sin(B_n)\sin(2\pi n f t_d + B_n)}{|t_F \sin^2(A_n) - t_R \sin(B_n)\sin(2\pi n f t_d + B_n)|} \right]$$

The same values of A_n, B_n, t_R, t_F, and t_d as those used to evaluate $|V_n|$ are used here, of course, yielding the following table of radian values for ϕ_n:

					n					
	1	2	3	4	5	6	7	8	9	10
ϕ_n	2.23	4.39	3.08	−0.41	2.65	−0.96	3.24	−0.21	2.75	−0.62

With these values of $|V_n|$ and ϕ_n, it can then be graphically verified that

$$v(t) \cong \sum_{n=1}^{10} |V_n| \cos(2\pi f_n t - \phi_n)$$

A comparison of the given time-domain waveform for $v(t)$ and the sum of the first 10 of its frequency-domain components is shown in Fig. 4-15b.

Based on these results, transitions back and forth between the time domain and the frequency domain are possible with little or no additional mathematics. This will be useful in subsequent discussions here, and it should also be useful for practicing engineers.

CHAPTER 5

PERIODIC CIRCUIT CURRENTS AND THEIR MEASURED RADIATIONS

5.1 INTRODUCTION

Given the frequency-domain descriptions of periodic voltages developed in the previous chapter, it is a simple matter to similarly describe the circuit currents caused by those voltages. From those descriptions, it is then a simple matter to characterize the sinusoidal radiations of the currents. And those characterizations clearly identify the physical variables that determine the amplitudes of the sinusoidal radiations. Thus, are the physical parameters that can be adjusted to control excessive circuit current radiations clearly identified. In other words, upon reaching this point in the analysis of the problem, clear pointers to possible solutions will have been found.

It should be clear by now that the word "circuit" is used here in a very strict sense. However, a few comments may be in order, to make sure there is no misunderstanding. It is quite common among electrical engineers and technicians to use the words "circuit" and "network" to mean the same thing. That is not being done here. A circuit, as the word is used here, is a *single, closed, path* followed by a current. A circuit current is assumed to have essentially the same amplitude over its entire path, but not the same phase. Parts of a circuit current's path may be shared with other currents, but no two circuit currents can share the same path in its entirety.

A network, on the other hand, is a collection of circuits—at least two, and often many more. Therefore, the radiations of a network are the sum of the radiations of the circuit currents that make up the network.

5.2   SINUSOIDAL COMPONENTS OF PERIODIC CIRCUIT CURRENTS

As discussed in the previous chapter, any periodic time-varying voltage can be described in the frequency domain as

$$v(t) = \overline{v(t)} + \sum_{n=1}^{\infty} |V_n| \cos(2\pi f_n t - \phi_n) \tag{5-1}$$

In that description

$$\overline{v(t)} = \text{the average value of } v(t)$$
$$|V_n| = \sqrt{a_n^2 + b_n^2}$$
$$\phi_n = \arctan(b_n/a_n) + \pi(1 - a_n/|a_n|)/2$$
$$a_n \text{ and } b_n = \text{the Fourier coefficients of } v(t)$$

When that voltage $v(t)$ is supplied to an electrical network, the circuits of the network will respond with currents having similar descriptions. Each resulting circuit current will have the general frequency-domain description

$$i(t) = \overline{i(t)} + \sum_{n=1}^{\infty} |I_n| \cos(2\pi f_n t - \gamma_n) \tag{5-2}$$

For any value of n, the summands in Eqs. 5-1 and 5-2 will be related to each other by Ohm's law. Suppose, for example, a circuit to which $v(t)$ is applied has, at the frequency f_n, the impedance $Z_n = R + X_n$, where R is the resistance of the circuit, and X_n is the reactance at that frequency. Then Ohm's law says that $i_n(t) = v_n(t)/Z_n$. More specifically, if

$$v_n(t) = |V_n| \cos(2\pi f_n t - \phi_n)$$

then

$$i_n(t) = \frac{|V_n|}{Z_n} \cos(2\pi f_n t - \phi_n)$$

$$= \frac{|V_n|}{|Z_n|} \cos(2\pi f_n t - \phi_n - \psi_n)$$

$$= |I_n| \cos(2\pi f_n t - \gamma_n) \tag{5-3}$$

where

$$|Z_n| = \sqrt{R^2 + X_n^2} \quad \text{and}$$

$$\psi_n = \arctan\left(\frac{X_n}{R}\right) \tag{5-4}$$

Thus, at each frequency f_n

$$|I_n| = \frac{|V_n|}{|Z_n|} \quad \text{and} \quad \gamma_n = \phi_n + \psi_n \tag{5-5}$$

where $|Z_n| = \sqrt{R^2 + X_n^2}$, $\psi_n = \arctan(X_n/R)$, and $|V_n|$ and ϕ_n have the definitions given with Eq. 5-1.

The average current will be

$$\overline{i(t)} = \frac{\overline{v(t)}}{R} \tag{5-6}$$

where R is the resistance or zero-frequency impedance of the circuit. Thus, if there is a capacitance in the circuit, then $R \rightarrow \infty$ and $\overline{i(t)} = 0$. Otherwise, $R < \infty$ and $\overline{i(t)} = \overline{v(t)}/R \neq 0$.

If the voltage $v(t)$ of Eq. 5-1 has one of the periodic waveforms discussed in the previous chapter, then its amplitude is

$$|V_n| = \frac{V_p T D_n}{n^2 \pi^2} \tag{5-7}$$

Then the current in any circuit to which $v(t)$ is applied, when the circuit has the impedance Z_n at the frequency f_n, will be

$$i(t) = \frac{\overline{v(t)}}{R} + \sum_{n=1}^{\infty} \frac{V_p T D_n}{n^2 \pi^2 |Z_n|} \cos(\omega_n t - \phi_n - \psi_n) \tag{5-8}$$

Here, as previously defined, V_p is the peak amplitude of $v(t)$, T is its period, and D_n is its transition factor.

Example 5-1 Suppose the periodic voltage $v(t)$, which has the time-domain characteristics illustrated in Fig. 5-1, is to be applied to one or more circuits. Before the sinusoidal components of any circuit currents caused by $v(t)$ can be examined, the sinusoidal components of $v(t)$ must be found. In the time domain, $v(t)$ has a linear rise $t_r = T/50$ and a linear fall $t_f = T/20$. Its peak-to-peak voltage $V_p = 10$ volts, and the time from the start of its rise to the start

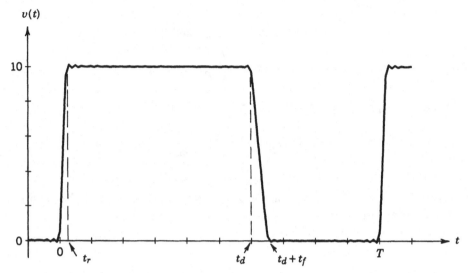

Figure 5-1. A time-varying voltage with period T and a linear rise and linear fall for which $t_r = T/50$, $t_f = T/20$, $t_d = 3T/5$, and $V_p = 10$ volts.

of its fall is $t_d = 3T/5$. When those values are substituted in Eq. 4-20 of the previous chapter, it follows that

$$a_n = -\frac{10}{n^2\pi^2}\left[50\sin^2\left(\frac{n\pi}{50}\right) - 20\sin\left(\frac{n\pi}{20}\right)\sin\left(\frac{5n\pi}{4}\right)\right]$$

and

$$b_n = \frac{10}{n^2\pi^2}\left[50\sin\left(\frac{n\pi}{50}\right)\cos\left(\frac{n\pi}{50}\right) - 20\sin\left(\frac{n\pi}{20}\right)\cos\left(\frac{5n\pi}{4}\right)\right] \quad (5\text{-}9)$$

Recalling that $|V_n| = \sqrt{a_n^2 + b_n^2}$, and $\phi_n = \arctan(b_n/a_n) + \pi(1 - a_n/|a_n|)/2$, values of $|V_n|$ and ϕ_n can be found for $v(t)$ for any frequency f_n. The first 10 values of $|V_n|$ and ϕ_n thus obtained are given in Table 5-1. The average value of this voltage is $\overline{v(t)} = 6.15$ volts.

Now, suppose this voltage is applied to a circuit with the schematic diagram of Fig. 5-2. Then the impedance seen by the voltage source at the frequency f_n is

$$Z_n = R + X_n = R - \frac{1}{\omega_n C} \quad (5\text{-}10)$$

so that

TABLE 5-1. Values of $|V_n|$ and ϕ_n for the Voltage $v(t)$ of Fig. 5-1

					n					
	1	2	3	4	5	6	7	8	9	10
$|V_n|$	5.94	2.09	0.97	1.52	0.28	0.81	0.66	0.19	0.58	0.24
ϕ_n	1.99	0.86	2.81	1.70	0.73	2.50	1.46	2.96	2.23	1.46

$$|Z_n| = \sqrt{R^2 + \left(-\frac{1}{\omega_n C}\right)^2} \quad \text{and} \quad \psi_n = \arctan\left(-\frac{1}{\omega_n CR}\right) \tag{5-11}$$

Therefore, if $R = 1000$ ohms, and the capacitance is such that the time constant $RC = T/20$, then $C = T/2 \times 10^{-4}$, and it follows that

$$|Z_n| = 1000\sqrt{1 + \left(\frac{10}{n\pi}\right)^2} \tag{5-12}$$

and

$$\psi_n = \arctan\left(-\frac{10}{n\pi}\right) \tag{5-13}$$

And because there is capacitance in the circuit, $\overline{i(t)} = 0$. Therefore, in the fre-

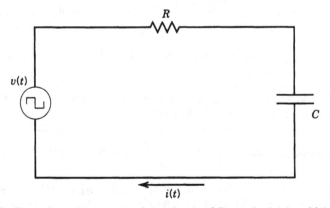

Figure 5-2. The schematic diagram of the circuit of Example 5-1 in which $R = 1000$ ohms and the values of C and T are such that the time constant RC equals $T/20$.

quency domain this circuit current is

$$i(t) = \sum_{n=1}^{\infty} |I_n| \cos(\omega_n t - \phi_n - \psi_n)$$

where

$$|I_n| = \frac{|V_n|/1000}{\sqrt{1 + (10/n\pi)^2}} \tag{5-14}$$

In evaluating these expressions, values of $|V_n|$ and ϕ_n for $n = 1$ to $n = 10$ can be taken from Table 5-1. Values of $|V_n|$ and ϕ_n for $n > 10$ can be found with values of a_n and b_n obtained with Eq. 5-9. And values of ψ_n can be found with Eq. 5-13. The first 10 values of $|I_n|$ and $\gamma_n = \phi_n + \psi_n$ for this current are given in Table 5-2. A time-domain plot of $i(t)$, which results from Eq. 5-14 and the first 40 values of $|I_n|$ and γ_n, is shown in Fig. 5-3.

Example 5-2   Next, suppose that the same voltage as that of the previous example is applied to a circuit that has the schematic diagram of Fig. 5-4. The impedance seen by the voltage source in this case will be

$$Z_n = R + \omega_n L$$

so that

$$|Z_n| = \sqrt{R^2 + (\omega_n L)^2} \quad \text{and} \quad \psi_n = \arctan\left(\frac{\omega_n L}{R}\right) \tag{5-15}$$

Then, in this case, if R is 1000 ohms and the time constant L/R is $T/20$, then

$$|Z_n| = 1000\sqrt{1 + (n\pi/10)^2}$$

TABLE 5-2. Values of $|I_n|$ and γ_n for the Current $i(t)$ in the Circuit of Fig. 5-2 with the Source Voltage of Fig. 5-1

					n					
	1	2	3	4	5	6	7	8	9	10
$|I_n|$	1.78	1.11	0.67	1.19	0.24	0.71	0.60	0.17	0.55	0.23
γ_n	0.73	−0.15	2.00	1.03	0.17	2.01	1.03	2.58	1.89	1.15

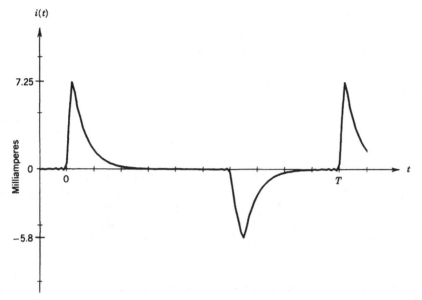

Figure 5-3. A time-domain representation of the circuit current $i(t)$ of Fig. 5-2 when the source voltage is that of Fig. 5-1.

and

$$\psi_n = \arctan\left(\frac{n\pi}{10}\right) \tag{5-16}$$

Because there is no capacitance in this circuit, the average current will be $\overline{i(t)} = \overline{v(t)}/R$. Therefore, the current will be

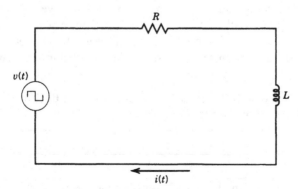

Figure 5-4. The schematic diagram of the circuit of Example 5-2 in which $R = 1000$ ohms and the values of L and T are such that the time constant L/R equals $T/20$.

$$i(t) = \frac{\overline{v(t)}}{R} + \sum_{n=1}^{\infty} |I_n| \cos(\omega_n t - \phi_n - \psi_n)$$

where

$$|I_n| = \frac{|V_n|/1000}{\sqrt{1 + (n\pi/10)^2}} \tag{5-17}$$

The values of $\overline{v(t)}$, $|V_n|$, and ϕ_n are all given in the previous example, and ψ_n is defined in Eq. 5-16.

Values of $|I_n|$ and γ_n for this current are given in Table 5-3 for the first 10 values of n. Also, $\overline{i(t)} = \overline{v(t)}/R = 6.15/1000 = 6.15$ ma. The time-domain representation of the current that is obtained with these results is illustrated in Fig. 5-5.

Example 5-3 Finally, consider a network that has the schematic diagram of Fig. 5-6, and suppose again that the applied voltage is that of Example 5-1. This network consists of two circuits: the path followed by the current $i_C(t)$ through the capacitance C, and the path followed by the current $i_2(t)$ through the resistance R_2. Where these paths coincide, the current is $i_C(t) + i_2(t) = i_1(t)$, the current through the resistance R_1.

A brief digression is appropriate here. In the previous paragraph, the words "capacitance" and "resistance" were used rather than "capacitor" and "resistor." The reason for this is that the lumped impedance components of a network may not always completely account for the true characteristics of a circuit, so far as radiations are concerned. For example, "stray" capacitance is often encountered in searching for causes of unintentional radiations. Also, when common ground connections are used to return a current to its source, more than one path may be followed by the current, creating more than one circuit. In other words, the physical implementation of a network can have a definite effect on its electrical characteristics. And any such effects may be significant so far as radiations are concerned. Therefore, whenever the electrical effects of a physical implementation are recognized, they should be included in schematic diagrams.

TABLE 5-3. Values of $|I_n|$ and ψ_n for the Current $i(t)$ in the Circuit of Fig. 5-4 with the Source Voltage of Fig. 5-1

	\multicolumn{10}{c}{n}
	1	2	3	4	5	6	7	8	9	10
$|I_n|$	5.67	1.77	0.71	0.95	0.15	0.38	0.27	0.07	0.19	0.07
γ_n	2.30	1.42	3.57	2.60	1.74	3.58	2.60	4.16	3.46	2.72

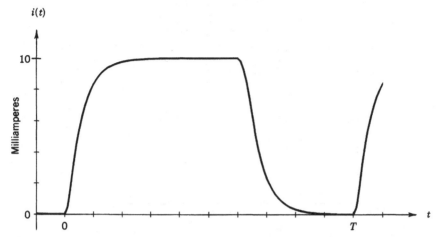

Figure 5-5. A time-domain representation of the circuit current $i(t)$ of Fig. 5-4 when the source voltage is that of Fig. 5-1.

If no such effects are recognized, it should be carefully verified that there are no such effects of any significance. This is discussed at more length in the next two chapters.

In obtaining the frequency-domain descriptions of the circuit currents in the network of this example it will be advantageous to first find the current $i_C(t)$, then it will be a simple matter to find the voltage $v(t)$ and the current $i_2(t) = v_C(t)/R_2$.

To find $i_C(t)$, a network will be used that is equivalent to that of Fig. 5-6 so far as the capacitance is concerned. The schematic of the equivalent network is illustrated in Fig. 5-7. The equivalence of the two networks, from the view-

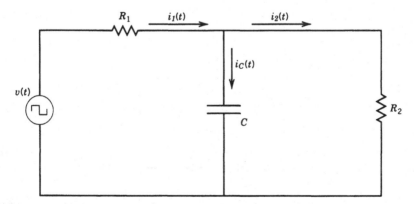

Figure 5-6. A network consisting of two circuits: the circuit of the current $i_C(t)$ and the circuit of the current $i_2(t)$.

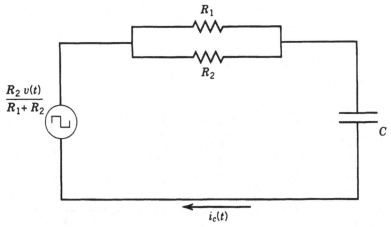

Figure 5-7. A network of circuits that is the equivalent of the network of Fig. 5-6 insofar as the capacitance C is concerned.

point of the capacitance C, can be seen as follows. If C is removed from either network, $i_C(t) = 0$ and the voltage where C had been connected will be equal to $v(t)R_2/(R_1+R_2)$. The source impedance seen by C will be $R_1R_2/(R_1+R_2)$, in either network. Therefore, so far as the capacitance C is concerned, the networks are the same. The current $i_C(t)$ and the voltage $v_C(t)$ will be the same in both networks.

In the network of Fig. 5-7, the load impedance seen by the source is

$$Z_n = \frac{R_1R_2}{R_1 + R_2} - \frac{1}{\omega_n C}$$

so that

$$|Z_n| = \sqrt{\left(\frac{R_1R_2}{R_1 + R_2}\right)^2 + \left(\frac{1}{\omega_n C}\right)^2} \quad \text{and}$$

$$\psi_n = \arctan\left(-\frac{R_1 + R_2}{\omega_n R_1 R_2 C}\right) \tag{5-18}$$

Therefore, if $R_1 = 1000$ ohms, $R_2 = 500$ ohms, and the time constant $R_1 C = T/5$, then it follows that

$$|Z_n| = \sqrt{\left(\frac{5000}{15}\right)^2 + \left(\frac{5000}{2\pi n}\right)^2}$$

$$= 1000\sqrt{\left(\frac{1}{3}\right)^2 + \left(\frac{5}{2\pi n}\right)^2} \qquad (5\text{-}19)$$

and

$$\psi_n = \arctan\left(-\frac{7.5}{n\pi}\right) \qquad (5\text{-}20)$$

Since the voltage source in the network of Fig. 5-7 is $R_2 v(t)/(R_1 + R_2) = v(t)/3$, it is seen that

$$|I_{nC}| = \frac{|V_n|}{3|Z_n|} = \frac{|V_n|/1000}{\sqrt{1 + (7.5/n\pi)^2}} \qquad (5\text{-}21)$$

And, of course, $\overline{i_C(t)} = 0$.

Thus, the current in the capacitance in either of the networks of Fig. 5-6 or Fig. 5-7 will be

$$i_C(t) = \sum_{n=1}^{\infty} |I_{nC}| \cos(\omega_n t - \phi_n - \psi_n) \qquad (5\text{-}22)$$

where $|I_{nC}|$ is given in Eq. 5-21, ψ_n is given in Eq. 5-20, and $|V_n|$ and ϕ_n have the values given for the voltage $v(t)$ of Example 5-1. The values of $|I_{nC}|$ and $\gamma_{nC} = \phi_n + \psi_n$ for $n = 1$ to $n = 10$, are given in Table 5-4, and the time-domain representation of $i_C(t)$ is illustrated in Fig. 5-8.

TABLE 5-4. Values of $|I_{nC}|$ (ma) and γ_{nC} (rad) for the Current $i_C(t)$ in Figs. 5-6 and 5-7

					n					
	1	2	3	4	5	6	7	8	9	10
$\lvert I_{nC}\rvert$	2.30	1.34	0.76	1.31	0.26	0.75	0.63	0.18	0.56	0.24
γ_{nC}	0.82	−0.02	2.14	1.61	0.29	2.12	1.13	2.67	1.97	1.22

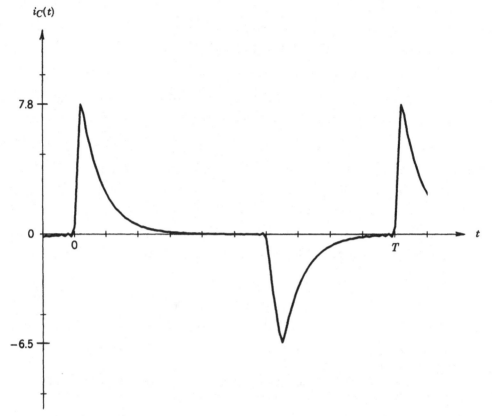

Figure 5-8. A time-domain representation of the current $i_C(t)$ in the capacitance of Figs. 5-6 and 5-7.

The other circuit current, $i_2(t)$, in the network of Fig. 5-6 can now be found based on the results obtained for $i_C(t)$. The voltage $v_C(t)$ across the capacitance C is also the voltage across the resistance R_2 in the network of Fig. 5-6. Therefore, the current through resistor R_2 is $i_2(t) = v_C(t)/R_2$. And, for any frequency f_n, the sinusoidal component of the current $i_C(t)$ will be $|I_{nC}| \cos(\omega_n t - \gamma_{nC})$. Therefore, since $-1/\omega_n C$ is the reactance of the capacitance at the frequency f_n, and $i_c(t)$ will lead $v_c(t)$ in phase by $\pi/2$ radians, the voltage $v_C(t)$ will have the frequency domain representation

$$v_C(t) = \sum_{n=1}^{\infty} \frac{|I_{nC}|}{\omega_n C} \cos\left(\omega_n t - \gamma_{nC} + \frac{\pi}{2}\right) \qquad (5\text{-}23)$$

And since $i_2(t) = v_C(t)/R_2$, and $\overline{i_2(t)} = \overline{v(t)}/(R_1 + R_2)$ in the network of Fig. 5-6, it follows that

$$i_2(t) = \frac{\overline{v(t)}}{R_1 + R_2} + \sum_{n=1}^{\infty} |I_{n2}| \cos(\omega_n t - \gamma_{n2})$$

where

$$|I_{n2}| = \frac{|I_{nC}|}{\omega_n R_2 C} \quad \text{and} \quad \gamma_{n2} = \gamma_{nC} - \frac{\pi}{2} \tag{5-24}$$

Now, since $R_1 = 1000$ ohms, $R_2 = 500$ ohms, $R_2 C = T/10$, and $|I_{nC}|$ is given in Eq. 5-21, it follows from Eq. 5-24 that

$$
\begin{aligned}
|I_{n2}| &= \frac{5|I_{nC}|}{n\pi} = \frac{5}{n\pi} \left(\frac{|V_n|/1000}{\sqrt{1 + (7.5/n\pi)^2}} \right) \\
&= \frac{|V_n|/1000}{\sqrt{(n\pi/5)^2 + (1.5)^2}}
\end{aligned}
$$

and

$$\gamma_{n2} = \phi_n + \psi_n - \frac{\pi}{2} \tag{5-25}$$

Here, again, $|V_n|$ and ϕ_n are the values obtained for $v(t)$ in Example 5-1, and ψ_n is given in Eq. 5-20. Values of $|I_{n2}|$ and γ_{n2} for $n = 1$ to $n = 10$ are given in Table 5-5. The time-domain representation of $i_2(t)$ obtained from these values is illustrated in Fig. 5-9.

It is now possible to examine the radiations of circuit currents such as those considered above.

TABLE 5-5. Values of $|I_{n2}|$ (ma) and γ_{n2} (rad) for the Current $i_2(t)$ in the Network of Fig. 5-6

					n					
	1	2	3	4	5	6	7	8	9	10
$|I_{n2}|$	3.65	1.07	0.40	0.52	0.08	0.20	0.14	0.04	0.10	0.04
γ_{n2}	2.39	1.55	3.71	2.73	1.86	3.69	2.70	4.24	3.54	2.79

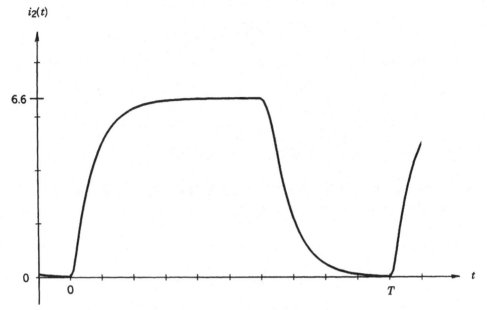

Figure 5-9. A time-domain representation of the current $i_2(t)$ in the network of Fig. 5-6.

5.3   SINUSOIDAL RADIATIONS OF PERIODIC CIRCUIT CURRENTS

The maximum radiated electric fields of a variety of rectangular circuits with sinusoidal currents were found in Chapter 3. It was seen there that short-length and medium-length rectangular circuits with the current $i(t) = |I| \sin(\omega t)$ cause maximum radiated electric fields

$$
\begin{aligned}
\max |E| &= \frac{Z_0 |I| f W}{cd} \; \sin\left[\frac{\pi f (2L + W)}{c} \right] \\
&= \frac{Z_0 |I| W_\lambda}{d} \; \sin[\pi(2L_\lambda + W_\lambda)]
\end{aligned}
\tag{5-26}
$$

It was also seen that long rectangular circuits with the current $i(t) = |I| \sin(\omega t)$ will cause maximum radiated electric fields

$$
\max |E| = \frac{Z_0 |I| f W}{cd} = \frac{Z_0 |I| W_\lambda}{d}
\tag{5-27}
$$

In these equations, $Z_0 = 120\pi$ ohms, the characteristic impedance of free space, $|I|$ is the peak amplitude of $i(t)$, f is the frequency of $i(t)$, L and W are the length and width of the rectangular path of $i(t)$, and d is the observation distance. Both Eqs. 5-26 and 5-27 apply to rectangular circuits of widths $W_\lambda = Wf/c \leq \frac{1}{16}$. Equation 5-26 applies to rectangular circuits of lengths $L_\lambda \leq \frac{1}{16}$, which are short-length circuits, and to circuits of lengths $\frac{1}{16} < L_\lambda = Lf/c \leq \frac{1}{4} - W_\lambda/2$, which are medium-length circuits. And, Eq. 5-27 applies to long circuits of lengths $L_\lambda > (\frac{1}{4} - W_\lambda/2)$. In these circuits, the source is centered in one current segment of length W and the load is assumed to be at or near the center of the other segment of length W. The position of the load is arbitrary, but the position of the source determines the phase relationships that have been assumed for the current segments.

Circuits that satisfy the dimensional requirements of Eq. 5-26 for short-length and medium-length circuits typically have maximum radiations in the direction from load to source. However, the maximum radiations of long circuits, described by Eq. 5-27, vary in direction with the length of the circuit and may occur in almost any direction. The direction of maximum radiation is relatively unimportant, however, because measurements must be made in almost every direction to verify regulatory compliance.

It is readily inferred from the above that the maximum radiated electric field of the nth sinusoidal component $|I_n| \sin(2\pi f_n t)$ of any periodic current $i(t)$ in a short-length or medium-length rectangular circuit will be

$$\max |E_n| = \frac{Z_0 |I_n| f_n W}{cd} \sin \left[\frac{\pi f_n (2L + W)}{c} \right]$$

And the same current in a long rectangular circuit will cause

$$\max |E_n| = \frac{Z_0 |I_n| f_n W}{cd} \tag{5-28}$$

These are the maximum electric fields that would be radiated in free space. However, as previously noted, unintentional electromagnetic radiations are measured in simulated half spaces. Therefore, the above equations must be modified, for the radiations modeled to be measured circuit current radiations.

5.4 MEASURED CIRCUIT CURRENT RADIATIONS

The sinusoidal components $E_n(t) = |E_n| \cos[2\pi f_n(t - d/c)]$ of the electric fields radiated by periodic circuit currents are often significant at numerous frequencies f_n. In measuring those components,

- The amplitude $|E_n|$ is measured at each frequency of occurrence f_n.
- Maximum measured values of $|E_n|$ are sought in all directions.

- The measurement environment causes a mirror image of $E_n(t)$, the radiation of which is included in the measured values of $|E_n|$.

Therefore, because maximum values are sought, the measured value of a periodic circuit current's radiation at any frequency f_n will be

$$\text{meas}|E_n| = 2(\max |E_n|) = 2 \, \frac{Z_0|I_n|f_n W}{cd} \, \sin\left[\frac{\pi f_n(2L + W)}{c}\right]$$

for short-length and medium-length rectangular circuits, and

$$\text{meas}|E_n| = 2(\max |E_n|) = 2 \, \frac{Z_0|I_n|f_n W}{cd} \tag{5-29}$$

for long circuits.

The sinusoidal components of periodic circuit currents that are caused by periodically varying voltages such as those considered in the previous chapter will have amplitudes

$$|I_n| = \frac{|V_n|}{|Z_n|} = \frac{V_p T}{n^2 \pi^2} \frac{D_n}{|Z_n|} \tag{5-30}$$

And since $f_n = nf$ and $f = 1/T$, it follows from Eqs. 5-29 and 5-30 that for short-length and medium-length circuits,

$$\text{meas}|E_n| = \frac{2Z_0 W}{n\pi^2 cd} \frac{V_p D_n}{|Z_n|} \, \sin\left[\frac{n\pi f(2L + W)}{c}\right] \tag{5-31}$$

and for long circuits,

$$\text{meas}|E_n| = \frac{2Z_0 W}{n\pi^2 cd} \frac{V_p D_n}{|Z_n|} \tag{5-32}$$

These equations clearly identify those parameters that are most important in reducing and controlling the radiated electric fields of the periodic circuit currents previously discussed.

More specifically, when a periodic source voltage $v(t)$ causes circuit currents $i(t)$, those currents cause radiated electric fields $E(t)$. And when $E(t)$ is measured in the frequency domain, its components are directly proportional to

V_p, the peak amplitude of $v(t)$

D_n, the transition factor of $v(t)$

W, the width of the area enclosed by the path of $i(t)$

Also, those measured radiations are inversely proportional to $|Z_n|$, the magnitude of a circuit's impedance. Both $|Z_n|$ and D_n vary with the frequency f_n, and D_n is also a function of the constant time-domain waveform parameters t_r, t_f, and t_d. Of the other parameters, V_p is constant for any given voltage waveform, and W is constant for a given current path. The variation of D_n is briefly examined in the next section, and the adjustment of V_p, D_n, W, and $|Z_n|$, is examined in detail in the next chapter.

The circuit length L is clearly not significant in controlling measured radiations. It has little or no effect, except at very low frequencies f_n, in circuits for which both $Lf_n/c \leq \frac{1}{16}$ and $Wf_n/c \leq \frac{1}{16}$, so that $\sin[\pi f_n(2L + W)/c] \cong \pi f_n(2L + W)/c$.

5.5 THE EFFECT OF THE TRANSITION FACTOR

The effect on the values of $\text{meas}|E_n|$ of the transition factor D_n is not quite as straightforward as the effects of the other factors in Eq. 5-31. However, once D_n is clearly understood, the seemingly erratic behavior of $\text{meas}|E_n|$ that is often seen in practice becomes quite understandable, and how D_n can be changed to reduce $\text{meas}|E_n|$ also becomes clear.

As defined in the previous chapter, the transition factor of the time-domain description of a periodic voltage $v(t)$ is

$$D_n = \sqrt{\frac{\sin^2(A_n)}{t_R^2} - 2\,\frac{\sin(A_n)\sin(B_n)}{t_R t_F}\cos(\omega_n t_d - A_n + B_n) + \frac{\sin^2(B_n)}{t_F^2}}$$

$$(5\text{-}33)$$

where $t_R = t_r$ and $A_n = \pi f_n t_r$, for a linear rise; $t_R = t_r/e$ and $A_n = \arctan(\pi f_n t_r/e)$ for an exponential rise; $t_F = t_f$ and $B_n = \pi f_n t_f$, for a linear fall; and $t_F = t_f/e$ and $B_n = \arctan(\pi f_n t_f/e)$ for an exponential fall. As also previously defined, t_r is the 100% rise time of $v(t)$, t_f is the 100% fall time of $v(t)$, and t_d is the time from the start of the rise to the start of the fall.

In the expression for D_n, notice that when $\cos(\omega_n t_d - A_n + B_n) = \pm 1$, the quantity under the square root sign in Eq. 5-33 is a perfect square. Therefore, since $-1 \leq \cos(\omega_n t_d - A_n + B_n) \leq +1$, it can be seen that

$$D_n \leq \frac{|\sin(A_n)|}{t_R} + \frac{|\sin(B_n)|}{t_F} \leq \frac{1}{t_R} + \frac{1}{t_F} \qquad (5\text{-}34)$$

In other words, the maximum value of D_n can be reduced by increasing either the rise time of $v(t)$ or the fall time, or both. Based on these relationships and Eqs. 5-31 and 5-32, it can be seen that for short-length and medium-length circuits

$$\text{meas}|E_n| \le \frac{2Z_0 W}{n\pi^2 cd} \frac{V_p}{|Z_n|} \sin(\pi f_n (2L + W)/c)$$

$$\cdot \left[\frac{|\sin(A_n)|}{t_R} + \frac{|\sin(B_n)|}{t_F} \right] \tag{5-35}$$

and for long circuits

$$\text{meas}|E_n| \le \frac{2Z_0 W}{n\pi^2 cd} \frac{V_p}{|Z_n|} \left[\frac{|\sin(A_n)|}{t_R} + \frac{|\sin(B_n)|}{t_F} \right] \tag{5-36}$$

The practical implications of these relationships are probably best introduced by example. The general procedure to follow in predicting and controlling the measured values of the radiated electric field of a given circuit and voltage consists of the following initial steps.

1. Based on applicable regulations, determine the *maximum allowable* values of $\text{meas}|E_n|$ at all frequencies of concern.
2. Based on the above results, determine the *maximum expected* values of $\text{meas}|E_n|$ at all frequencies of concern.
3. If maximum expected values exceed maximum allowable values at any frequencies, then determine the *exact* values of $\text{meas}|E_n|$ to expect at those frequencies.
4. Repeat steps 2 and 3 as necessary.

This procedure is implemented in the example that follows.

Example 5-4 Suppose that the radiated emissions limits that are not to be exceeded are the combined Class B limits of CISPR, VDE, and the FCC from 30 to 1000 MHz.[1] The combined limits are the result of taking the lowest of the limits of these agencies at each frequency in that range. The resulting maximum allowable levels of electric field strength at those frequencies at a measurement distance of 10 meters from the source are shown in Fig. 5-10. These are the

---

[1]CISPR (Comite International Special des Perturbations Radioelectriques), the International Special Committee on Radio Frequency Interference, was established in 1934 by the International Electrotechnical Commission (IEC) to help promote international agreement in solving radio interference problems. VDE (Verband Deutscher Elektrotechniker), the Association of German Electrical Engineers, has been verifying compliance with radiated emissions limits since those limits were first established in Germany in 1949. The FCC, the United States Federal Communications Commission, began to enforce control on radiated emissions from computing devices in 1980, broadening that to include all digital devices in 1989.

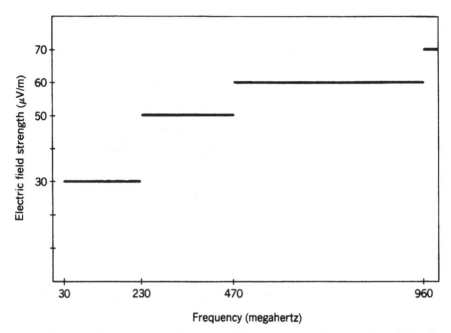

Figure 5-10. Combined CISPR, VDE, and FCC Class B radiated emissions limits for the 10-meter measurement distance ($d = 10$ m).

levels that $\text{meas}|E_n|$ is not to exceed at any frequency from 30 MHz to 1000 MHz.

Suppose the circuit and source voltage are those of Example 5-1. The circuit impedance has a magnitude of $|Z_n| = 1000\sqrt{1 + (10/n\pi)^2}$ ohms, and the time-domain voltage parameters are $V_p = 10$ volts, $t_r = T/50$, $t_d = 3T/5$, and $t_f = T/20$. Also suppose that the frequency $f = 20$ MHz and the circuit's width and length are $W = 1$ cm and $L = 4.5$ cm. Then $W_\lambda \leq \frac{1}{30}$ at 1000 MHz and $L_\lambda = \frac{4.5}{30} = 0.15$ at 1000 MHz, so that $L_\lambda < (\frac{1}{4} - W_\lambda/2)$, and this circuit is of short or medium length at all frequencies of concern. Finally, as specified in Fig. 5-10, the measurement distance will be $d = 10$ meters.

The next step is to determine the maximum values of $\text{meas}|E_n|$ to expect. Given that $t_r = T/50$ and $t_f = T/20$, and the voltage has a linear rise and fall, it follows that $A_n = n\pi/50$, $B_n = n\pi/20$, and

$$D_n \leq \frac{50}{T} \left| \sin\left(\frac{n\pi}{50}\right) \right| + \frac{20}{T} \left| \sin\left(\frac{n\pi}{20}\right) \right| \tag{5-37}$$

Therefore, with $V_p = 10$ volts, $f = 20$ MHz, $f/c = \frac{1}{15}$, $W = 1$ cm, $L = 4.5$ cm, $|Z_n| = 1000\sqrt{1 + (10/n\pi)^2}$ ohms, and $d = 10$ meters,

$$\text{meas}|E_n| = \frac{2Z_0W}{n\pi^2cd} \frac{V_pD_n}{|Z_n|} \sin\left[\frac{\pi f_n(2L+W)}{c}\right]$$

$$= \frac{2Z_0}{n\pi^2} \frac{W}{cd} \frac{V_pD_n}{|Z_n|} \sin\left[\frac{n\pi(2L+W)f}{c}\right]$$

$$= \frac{240}{n\pi} \frac{10^{-2}}{3\times10^9} \frac{10D_n}{1000} \frac{\sin[n\pi(10^{-1})/15]}{\sqrt{1+(10/n\pi)^2}}$$

$$= \frac{8\times10^{-6}D_n}{n\pi\sqrt{1+(10/n\pi)^2}} \sin\left(\frac{n\pi}{150}\right) \frac{\mu\,\text{Volts}}{\text{meter}} \qquad (5\text{-}38)$$

From Eqs. 5-37 and 5-38, then, since $1/T = f = 20$ MHz, it follows that

$$\text{meas}|E_n| \leq \frac{8\times10^{-6}[(50/T)|\sin(n\pi/50)| + (20/T)|\sin(n\pi/20)|]}{n\pi\sqrt{1+(10/n\pi)^2}} \sin\left(\frac{n\pi}{150}\right)$$

$$= \frac{160[50|\sin(n\pi/50)| + 20|\sin(n\pi/20)|]}{n\pi\sqrt{1+(10/n\pi)^2}} \sin\left(\frac{n\pi}{150}\right)$$

$$= \frac{3200[2.5|\sin(n\pi/50)| + |\sin(n\pi/20)|]}{n\pi\sqrt{1+(10/n\pi)^2}} \sin\left(\frac{n\pi}{150}\right) \frac{\mu\,\text{Volts}}{\text{meter}}$$

$$(5\text{-}39)$$

These maximum values of meas$|E_n|$ for this circuit and voltage are plotted and compared to the combined limits of CISPR, VDE, and the FCC in Fig. 5-11. The plot shows that meas$|E_n|$ may equal or exceed regulatory limits at all frequencies from about 100 to 680 MHz. Therefore, something clearly needs to be done to make this circuit less radiative. However, for added certainty, meas$|E_n|$ will be calculated exactly.

From Eq. 5-33, with $1/T = f = 20$ MHz, it is seen that

$$D_n = \frac{20}{T} \sqrt{6.25\sin^2\left(\frac{n\pi}{50}\right) - 5\sin\left(\frac{n\pi}{50}\right)\sin\left(\frac{n\pi}{20}\right)\cos\left(\frac{123n\pi}{100}\right) + \sin^2\left(\frac{n\pi}{20}\right)}$$

$$= 400\sqrt{6.25\sin^2\left(\frac{n\pi}{50}\right) - 5\sin\left(\frac{n\pi}{50}\right)\sin\left(\frac{n\pi}{20}\right)\cos\left(\frac{123n\pi}{100}\right) + \sin^2\left(\frac{n\pi}{20}\right)} \quad \text{MHz}$$

so that

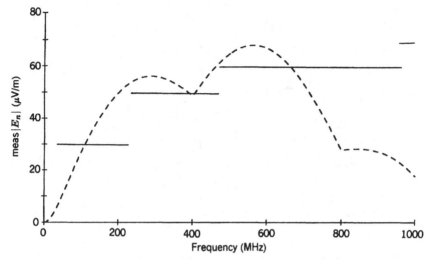

Figure 5-11. Maximum expected values of meas$|E_n|$ (- - -) versus combined regulatory limits (—), at the 10-meter measurement distance ($d = 10$ m).

$$\text{meas}|E_n| = \frac{3200\sqrt{6.25 \sin^2\left(\frac{n\pi}{50}\right) - 5 \sin\left(\frac{n\pi}{50}\right) \sin\left(\frac{n\pi}{20}\right) \cos\left(\frac{123n\pi}{100}\right) + \sin^2\left(\frac{n\pi}{20}\right)}}{n\pi\sqrt{1 + (10/n\pi)^2}}$$

μV/m   (5-40)

Values of meas$|E_n|$ obtained with these equations are shown in Fig. 5-12. There are 11 values of meas$|E_n|$ that equal or exceed the combined regulatory limits. Therefore, this voltage and circuit are clearly unacceptable as they are currently proposed.

Since this circuit and voltage are unacceptable, they need to be modified and the radiations need to be reexamined. Suppose, therefore, that the circuit's length must remain at $L = 4.5$ cm, but its width can be reduced to $W = 0.5$ cm. Then, Eq. 5-38 becomes

$$\text{meas}|E_n| = \frac{4 \times 10^{-6} D_n}{n\pi\sqrt{1 + (10/n\pi)^2}} \sin\left(\frac{0.95n\pi}{150}\right) \quad \mu\text{V/m} \qquad (5\text{-}41)$$

but the maximum value of D_n remains the same as in Eq. 5-37. Thus, the maximum value of meas$|E_n|$ will be slightly less than one-half its previous value. The result is shown in Fig. 5-13. The radiated emissions from this circuit and voltage can now be expected to fully comply with regulatory limits at all frequencies from 30 to 1000 MHz.

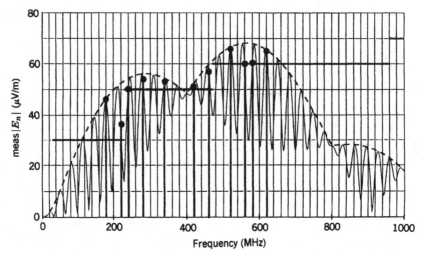

Figure 5-12. Expected values of meas$|E_n|$ (•) from the circuit of Example 5-4 compared to their expected maximum values (- - -) and the combined regulatory limits (—).

Finally, it is interesting to observe why, in practice, measured values of $|E_n|$ vary up and down in a seemingly erratic manner from one frequency f_n to the next. As noted in developing Eqs. 5-33, the variations occur because of the term $\cos(\omega_n t_d - A_n + B_n)$ in the transition factor D_n. To better see this, consider Fig. 5-14. In that figure, meas$|E_n|$ for the original circuit and voltage of the previous example is plotted as a continuous function of frequency from 200 to 600 MHz. Also plotted and circled are the values of meas$|E_n|$ that would be

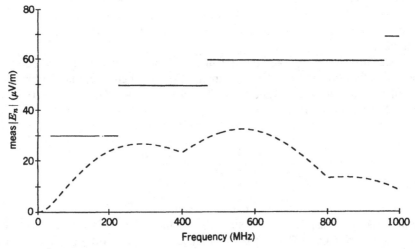

Figure 5-13. Maximum meas$|E_n|$ (- - -) expected from the modified circuit of Example 5-4 compared to the combined regulatory limits (—).

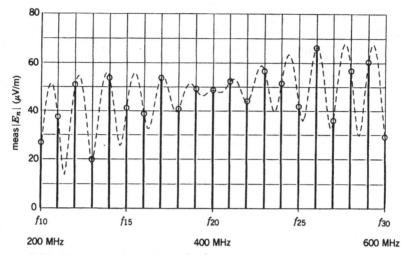

Figure 5-14. Observed values of meas$|E_n|$ (o) appear to fluctuate erratically because the transition factor D_n varies in a sinusoidal fashion with frequency.

measured at frequencies from f_{10} to f_{30}. From these plots, it is that seen that measured values of $|E_n|$ do jump around from one frequency f_n to the next, but the behavior is quite predictable.

5.7 SUMMARY

As seen in Chapter 4, any periodic, time-varying voltage $v(t)$ of frequency f has the frequency-domain description

$$v(t) = \overline{v(t)} + \sum_{n=1}^{\infty} |V_n| \cos(2\pi f_n t - \phi_n) \qquad (5\text{-}42)$$

In other words, the voltage $v(t)$ is the sum of its average value $\overline{v(t)}$ and an infinite number of sinusoidal components, $|V_n| \cos(2\pi f_n t - \phi_n)$. And, if Z is the impedance of the circuit for which $v(t)$ is the source voltage, then the resulting circuit current $i(t)$ will be

$$i(t) = \frac{v(t)}{Z}$$

$$= \frac{\overline{v(t)}}{Z} + \frac{1}{Z} \sum_{n=1}^{\infty} |V_n| \cos(2\pi f_n t - \phi_n)$$

$$= \frac{\overline{v(t)}}{R} + \sum_{n=1}^{\infty} \frac{|V_n|}{Z_n} \cos(2\pi f_n t - \phi_n) \qquad (5\text{-}43)$$

The impedance $Z = R$, a resistance, when the frequency is zero; therefore, the average current is $\overline{i(t)} = \overline{v(t)}/R$. In general, however, the impedance Z will be a function of frequency which exhibits different values of impedance for different frequencies f_n. Thus, the notation $Z = Z_n$ is used when the frequency is $f_n = nf$. In other words, when $n \geq 1$, the frequency f_n will be greater than zero, and the sinusoidal current components will be

$$
\begin{aligned}
i_n(t) &= \frac{|V_n|}{Z_n} \cos(2\pi f_n t - \phi_n) \\
&= \frac{|V_n|}{|Z_n|} \cos(2\pi f_n t - \phi_n + \psi_n) \\
&= |I_n| \cos(2\pi f_n t - \gamma_n)
\end{aligned}
\tag{5-44}
$$

The angle ψ_n is the phase difference between the sinusoidal components of voltage and current that is caused by Z_n when the frequency is f_n.

Thus, any circuit current $i(t)$ that is caused by the periodic, time-varying voltage $v(t)$ of Eq. 5-42 has the frequency-domain description

$$
i(t) = \overline{i(t)} + \sum_{n=1}^{\infty} |I_n| \cos(2\pi f_n t - \gamma_n)
\tag{5-45}
$$

This current is the sum of its average value $\overline{i(t)}$ and an infinite number of sinusoidal current components $|I_n| \cos(2\pi f_n t - \gamma_n)$.

It was determined in Chapter 3 what the maximum radiated electric fields of sinusoidal currents in rectangular circuits will be. A short-length to medium-length rectangular circuit with component currents $|I_n| \cos(2\pi f_n t - \gamma_n)$ will have maximum radiated electric field components

$$
\max |E_n| = \frac{Z_0 |I_n| f_n W}{cd} \sin\left[\frac{\pi f_n (2L + W)}{c} \right]
\tag{5-46}
$$

The same component currents in a long rectangular circuit will cause the maximum radiated electric field components

$$
\max |E_n| = \frac{Z_0 |I_n| f_n W}{cd}
\tag{5-47}
$$

To see why circuit-current radiations are a significant problem, consider the value of $\max |E_n|$ when $|I_n| = 1$ milliampere, $f_n = 100$ MHz, $W = 1$ centimeter, and $d = 10$ meters. Since $Z_0 = 120\pi$ ohms and $c = 3 \times 10^8$ meters/second, max $|E_n| = 40\pi \times 10^{-6} = 125.7$ microvolts/meter, which is more than twice the regulatory limit at 100 MHz.

In addition, the measured values of the electric fields radiated will be very close to twice these maximum values. This doubling occurs because of the measurement environment and the measurement procedures specified by regulations. Therefore, the measured electric fields radiated by any short-length or medium-length circuit current $i(t)$ will be

$$\text{meas}|E_n| \cong 2 \max |E_n|$$

$$= 2 \frac{Z_0|I_n|f_nW}{cd} \sin\left[\frac{\pi f_n(2L + W)}{c}\right] \qquad (5\text{-}48)$$

and those of a long circuit current will be

$$\text{meas}|E_n| \cong 2 \max |E_n|$$

$$= 2 \frac{Z_0|I_n|f_nW}{cd} \qquad (5\text{-}49)$$

Therefore, for the typical values of $|I_n|$, f_n, W, and d given in the example above, $\text{meas}|E_n|$ will be more than four times the regulatory limit.

The sinusoidal components of the voltages considered in the previous chapter have amplitudes

$$|V_n| = \frac{V_pTD_n}{n^2\pi^2} \qquad (5\text{-}50)$$

Therefore, when any such voltage is responsible for the current $i(t)$, the sinusoidal component currents have amplitudes

$$|I_n| = \frac{|V_n|}{|Z_n|} = \frac{V_pTD_n}{n^2\pi^2|Z_n|} \qquad (5\text{-}51)$$

And, finally, as shown in the present chapter,

$$D_n \leq \frac{|\sin(A_n)|}{t_R} + \frac{|\sin(B_n)|}{t_F} \leq \frac{1}{t_R} + \frac{1}{t_F} \qquad (5\text{-}52)$$

This equation shows that the maximum value of the transition factor equals the sum of the reciprocals of the rise and fall times of $v(t)$.

The following relationships perhaps best summarize and characterize all of the above. For short-length and medium-length circuits

$$\text{meas}|E_n| \le \frac{2Z_0 W}{n\pi^2 cd} \frac{V_p}{|Z_n|} \sin\left[\frac{\pi f_n(2L+W)}{c}\right]$$

$$\cdot \left[\frac{|\sin(A_n)|}{t_R} + \frac{|\sin(B_n)|}{t_F}\right] \tag{5-53}$$

And for long circuits

$$\text{meas}|E_n| \le \frac{2Z_0 W}{n\pi^2 cd} \frac{V_p}{|Z_n|} \left[\frac{|\sin(A_n)|}{t_R} + \frac{|\sin(B_n)|}{t_F}\right] \tag{5-54}$$

To relate the values of the variables in this expression to regulatory limits, suppose $W = 1$ cm, $V_p = 5$ volts, $|Z_n| = 1000$ ohms, $d = 10$ meters, $n = 10$, and $t_R = t_F = 2$ nanoseconds. Then, $Z_0/\pi^2 c = 40/\pi \times 10^{-8}$, $2WV_p/nd|Z_n| = 10^{-6}$, and $|\sin(A_n)|/t_R + |\sin(B_n)|/t_F \le 10^9$, so that meas$|E_n| \le 400/\pi \times 10^{-6} = 133.3$ μV/m, which exceeds the limits by a considerable amount at all frequencies.

These relationships say that to minimize the measured radiations of circuit currents established by source voltages $v(t)$ with periodic waveforms,

> Maximize t_R, the rise time of $v(t)$
> t_F, the fall time of $v(t)$
> $|Z_n|$, the impedance of the circuit
>
> Minimize V_p, the amplitude of $v(t)$
> W, the width of the circuital path of $i(t)$

Note that it is t_R and t_F that must be maximized to reduce D_n, and meas$|E_n|$. Increasing T or reducing f will only reduce meas$|E_n|$, if, in doing so, t_R and t_F are increased.

The parameters t_R, t_F, $|Z_n|$, V_p, and W are those with which the unintentional electromagnetic radiations of circuit currents must be controlled. However, the values of all of these parameters, except for W, are chosen primarily because of the primary functions the circuit currents are to perform. Therefore, the parameters other than W can seldom be adjusted to any great extent. For that reason, circuit width W is the key to effective reduction and control of unwanted radiations. This is discussed in much greater detail in the next chapter.

CHAPTER 6

CONTROL OF CIRCUIT-CURRENT RADIATIONS

6.1 INTRODUCTION

To control unwanted radiations from circuit currents, it is clear that $\text{meas}|E_n|$ must be minimized. The individual parameters that must be adjusted are clearly identified in the expressions given for $\text{meas}|E_n|$.

As previously determined, values of $\text{meas}|E_n|$ to be expected from small, medium-length, and long-length rectangular circuit currents can be calculated with the following equations. For small and medium-length circuits with source voltages such as those discussed in Chapters 4 and 5

$$
\begin{aligned}
\text{meas}|E_n| &= \left(\frac{2Z_0}{n\pi^2 cd} \right) \frac{V_p D_n}{|Z_n|} \, W \sin\left[\frac{n\pi f(2L + W)}{c} \right] \\
&= \frac{0.08}{n\pi} \left(\frac{V_p D_n}{|Z_n|} \right) W \sin\left[\frac{n\pi f(2L + W)}{c} \right] \quad \mu\text{V/m} \quad \text{(6-1a)}
\end{aligned}
$$

and for long circuits with those same source voltages,

$$
\begin{aligned}
\text{meas}|E_n| &= \left(\frac{2Z_0}{n\pi^2 cd} \right) \frac{V_p D_n W}{|Z_n|} \\
&= \frac{0.08}{n\pi} \left(\frac{V_p D_n W}{|Z_n|} \right) \quad \mu\text{V/m} \quad \text{(6-1b)}
\end{aligned}
$$

The second of these equations, in each case, results because $Z_0 = 120\pi$ ohms, $c = 300 \times 10^6$ meters/second, and the assumed measurement distance is $d = 10$ meters. The remaining parameters with which meas$|E_n|$ can be reduced are

V_p, the amplitude of $v(t)$, in volts

D_n, the transition factor of $v(t)$, in megahertz

Z_n, the impedance of the circuit, in ohms

W, the width of the circuit, in centimeters

These four parameters, V_p, D_n, Z_n, and W, and how they can be adjusted to minimize meas$|E_n|$ are the primary topics of discussion in this chapter. Each of these parameters is examined in order of increasing importance based on the extent to which each is usually adjustable. The order of discussion is V_p, Z_n, D_n, and W. In other words, to minimize circuit-current radiations, the source voltage amplitude V_p can seldom be changed significantly, whereas the circuit width W can almost always be reduced significantly. In fact, in a majority of cases, only W can be used to reduce circuit-current radiations. Nevertheless, reductions in W alone are usually quite sufficient to bring about more than adequate control of unwanted radiations.

The parameters L and f are of little use in controlling circuit-current radiations. It was seen in Chapter 3 that L, the length of a circuit, has little or no effect on the maximum radiation of a circuit current. And although D_n is a function of the frequency f of the voltage $v(t)$, by itself f has little or no effect on the magnitude of D_n. In summary, then, the primary subjects of discussion in this chapter are the proportionality

$$\text{meas}|E_n| \propto \frac{V_p D_n W}{|Z_n|} \tag{6-2}$$

and how V_p, Z_n, D_n, and W can be used to implement reductions in meas$|E_n|$.

6.2   SOURCE VOLTAGE AMPLITUDE

The voltage waveforms that were first discussed in Chapter 4 very often cause circuit currents that radiate excessively. Those voltage waveforms are usually found in digital electronic equipment. They receive primary attention here, because of their general tendency to cause unwanted electromagnetic radiation and because of their increasingly widespread use in modern electronic equipment.

In Eq. 6-1 it is seen that at every measurement frequency f_n, a circuit's radiated electric field is directly proportional to the amplitude V_p of its source voltage waveform $v(t)$. Consequently, at every measurement frequency $f_n = nf$, the value of meas$|E_n|$ will be reduced in direct proportion to any reductions that can be made to V_p. Unfortunately, from the point of view of controlling circuit radiations, V_p can seldom be reduced sufficiently to bring about any significant reduction in meas$|E_n|$ and much of the time it is irreducible. Values assigned to V_p are generally predetermined by the operations a circuit is designed to perform and by the technology with which it is implemented. Changes in V_p, no matter how slight, will often cause unacceptable changes in a circuit's performance. Thus, it is seldom possible to reduce meas$|E_n|$ significantly, if at all, by reducing V_p.

Based on these observations, it should be clear that V_p should be carefully and thoughtfully chosen in the initial stages of the design of a circuit. And thought should be given not only to what a circuit is intended to do, but also to what it should not do. To minimize and control circuit-current radiations, the amplitude of a circuit's source voltage should never be any larger than absolutely necessary.

6.3 CIRCUIT-CURRENT IMPEDANCE

It should not be surprising that maximizing the impedance a circuit presents to its currents tends to minimize its electromagnetic radiation. Maximizing impedance minimizes current, and time-varying currents are the cause of electromagnetic radiations. Therefore, it should be clear intuitively as well as mathematically that meas$|E_n|$ is inversely proportional to $|Z_n|$, the magnitude of a circuit's impedance. Thus, to whatever extent it may be possible, $|Z_n|$ should always be made as large as possible.

Since

$$|Z_n| = \sqrt{R^2 + X_n^2}$$

a circuit's resistance R and its reactance X_n should both be made large. The inductance L has the reactance

$$X_n = 2\pi f_n L = \omega_n L$$

and the capacitance C has the reactance

$$X_n = \frac{1}{2\pi f_n C} = \frac{1}{\omega_n C}$$

When both L and C are in a circuit together, their series reactance is

$$X_n = \omega_n L - \frac{1}{\omega_n C} = \frac{\omega_n^2 LC - 1}{\omega_n C}$$

Thus, to maximize $|Z_n|$, both R and L should always be made as large as possible, and C should always be made as small as possible. However, as already noted for V_p, the reasons for choosing particular values for R, L, and C go well beyond the reduction and control of unintentional radiation. All of the reasons for choosing their values must each be given appropriate attention. Nevertheless, to whatever extent it can be done, to minimize circuit-current radiation, $|Z_n|$ should always be made as large as possible.

Parallel reactance is not included in the above discussion, because parallel reactors reside in separate circuits and they must be examined separately. In other words, as a circuit is defined here, it has only *series* components.

6.4   SOURCE VOLTAGE WAVESHAPE

The transition factor D_n is a function of the shape of a periodic voltage waveform $v(t)$. Thus, changing the shape of the source voltage waveform of a circuit will change the magnitude of D_n, and, since meas$|E_n|$ is proportional to D_n, it will also change the magnitude of meas$|E_n|$. Several examples of the effects on meas$|E_n|$ of differently shaped source voltage waveforms are given below.

D_n is a function of the transitions of a periodic voltage $v(t)$ from one level to the other. The values of D_n depend on the transitions being linear, or exponential, on the times they require for completion, and on the time that elapses between them. As previously defined, those times are t_r, the 100% rise time; t_f, the 100% fall time; and t_d, the time from the start of t_r to the start of t_f. D_n is also a function of the frequencies $f_n = nf = n/T$, where $n = 1, 2, 3, \ldots \infty$. These are the frequencies of the sinusoidal components of $v(t)$, and the measurement frequencies—the frequencies at which meas$|E_n|$ is observed in practice.

Given these parameters of the waveform $v(t)$, its *transition factor D_n* can be described mathematically as

$$D_n = \sqrt{\frac{\sin^2(A_n)}{t_R^2} - 2\,\frac{\sin(A_n)}{t_R}\,\frac{\sin(B_n)}{t_F}\,\cos(2\pi f_n t_d - A_n + B_n) + \frac{\sin^2(B_n)}{t_F^2}}$$

$$(6\text{-}3)$$

In this expression, noting that $e = 2.718\ldots$,

- When the rise of $v(t)$ is linear, $t_R = t_r$ and $A_n = \pi f_n t_R$, and when the rise is exponential, $t_R = t_r/e$ and $A_n = \arctan(\pi f_n t_R)$.
- When the fall of $v(t)$ is linear, $t_F = t_f$ and $B_n = \pi f_n t_F$, and when the fall is exponential $t_F = t_f/e$ and $B_n = \arctan(\pi f_n t_F)$.

It can be seen from Eq. 6-3 that, whenever $\cos(\omega_n t_d - A_n + B_n) = \pm 1$,

$$D_n = \sqrt{\left[\frac{\sin(A_n)}{t_R} \mp \frac{\sin(B_n)}{t_F}\right]^2}$$

and

$$D_n \leq \frac{|\sin(A_n)|}{t_R} + \frac{|\sin(B_n)|}{t_F} \leq \frac{1}{t_R} + \frac{1}{t_F} \tag{6-4}$$

Thus, for small and medium-length circuits,

$$\text{meas}|E_n| \leq \left(\frac{2Z_0}{n\pi^2 cd}\right) \frac{V_p W}{|Z_n|} \left[\frac{|\sin(A_n)|}{t_R} + \frac{|\sin(B_n)|}{t_F}\right] \sin\left[\frac{n\pi f(2L+W)}{c}\right]$$

$$= \frac{0.08}{n\pi} \frac{V_p W}{|Z_n|} \left[\frac{|\sin(A_n)|}{t_R} + \frac{|\sin(B_n)|}{t_F}\right] \sin\left[\frac{n\pi f(2L+W)}{c}\right]$$

$$\equiv \text{Max}(\text{meas}|E_n|) \quad \mu V/m \tag{6-5a}$$

and for long circuits,

$$\text{meas}|E_n| \leq \left(\frac{2Z_0}{n\pi^2 cd}\right) \frac{V_p W}{|Z_n|} \left[\frac{|\sin(A_n)|}{t_R} + \frac{|\sin(B_n)|}{t_F}\right]$$

$$= \frac{0.08}{n\pi} \frac{V_p W}{|Z_n|} \left[\frac{|\sin(A_n)|}{t_R} + \frac{|\sin(B_n)|}{t_F}\right]$$

$$\equiv \text{Max}(\text{meas}|E_n|) \quad \mu V/m \tag{6-5b}$$

These expressions for $\text{Max}(\text{meas}|E_n|)$ yield the maximum possible value that $\text{meas}|E_n|$ can have at any given frequency f_n. From these expressions, it is quite apparent that the maximum values $\text{meas}|E_n|$ can reach can be reduced by increasing either t_R or t_F or both. Of course, the greatest reduction in $\text{Max}(\text{meas}|E_n|)$ will be achieved if both t_R and t_F are increased. It should be noted, however, that a reduction of $\text{Max}(\text{meas}|E_n|)$ does not necessarily imply a reduction in $\text{meas}|E_n|$, at all frequencies. Not all values of $\text{meas}|E_n|$ attain their maximum; therefore, at a few individual frequencies f_n, a reduction in $\text{Max}(\text{meas}|E_n|)$ may cause $\text{meas}|E_n|$ to increase. Nevertheless, when $\text{Max}(\text{meas}|E_n|)$ is reduced, at most frequencies $\text{meas}|E_n|$ will be reduced, and any increases in $\text{meas}|E_n|$ will be few and relatively small.

Several examples are now considered to illustrate the adjustment of D_n for the purpose of reducing $\text{meas}|E_n|$. However, as for V_p and Z_n, note that significant adjustment to D_n will seldom be an option. In practice, the primary reasons for D_n being the value it is will generally prevent any changes in it to reduce radiated emissions. Nevertheless, as also noted for V_p, and Z_n, whatever can be done to D_n to minimize $\text{meas}|E_n|$, should be done.

Example 6-1   The simple circuit of Fig. 6-1 will be the circuit in all of the examples that follow. It has a constant impedance of $Z_n = R = 1000$ ohms, a length $L = 5$ cm, and a width $W = 1$ cm. At $f_n \le 1000$ MHz, then, this is a medium-length circuit with a width $W_\lambda = Wf/c = \frac{1}{30}$, and length $\frac{1}{16} < L_\lambda = \frac{5}{30} = \frac{1}{6} < (\frac{1}{4} - W_\lambda/2) = \frac{7}{30}$.

A number of different source voltage waveshapes will be considered for this circuit, to compare the differences in values of $\text{meas}|E_n|$ that they cause. The voltages will all have different wave shapes, but otherwise they will be the same. They each have the same peak amplitude, $V_p = 10$ volts, the same frequency, $f = 40$ MHz, and the same period $T = 1/f = 25$ ns (nanoseconds). Therefore, the radiated electric field that will be measured from this circuit, at any frequency f_n from 30 MHz to 1000 MHz, at a measurement distance of $d = 10$ meters, will be

$$\text{meas}|E_n| = \frac{0.08}{n\pi} \frac{V_p D_n}{|Z_n|} W \sin\left[\frac{n\pi(2L+W)f}{c}\right]$$

$$= \frac{8 \times 10^{-6}}{n\pi} D_n \sin\left(\frac{11n\pi}{750}\right) \quad \mu V/m \tag{6-6}$$

The initial source voltage to be considered has the waveform illustrated in Fig. 6-2a. As noted above, that voltage has the frequency $f = 40$ MHz and the period $T = 25$ ns. It has a rise time of $t_r = T/25 = 1$ ns, a fall time $t_F = T/10 = 2.5$ ns, and a rise-to-fall time $t_d = T/2 = 12.5$ ns. Both of its transitions are linear, so that $t_R = t_r = T/25$, $A_n = n\pi/25$, $t_F = t_f = T/10$, and $B_n = n\pi/10$. Thus, the transition factor of this voltage waveform is

$$D_n = \sqrt{625 \frac{\sin^2(n\pi/25))}{T^2} - 500 \frac{\sin(n\pi/25)\sin(n\pi/10)}{T^2} \cos(1.06n\pi) + 100 \frac{\sin^2(n\pi/10)}{T^2}}$$

$$= 400\sqrt{6.25 \sin^2(n\pi/25) - 5\sin(n\pi/25)\sin(n\pi/10)\cos(1.06n\pi) + \sin^2(n\pi/10)} \quad \text{MHz} \tag{6-7}$$

From this expression it is seen that

$$D_n \le 1000\left|\sin\left(\frac{n\pi}{25}\right)\right| + 400\left|\sin\left(\frac{n\pi}{10}\right)\right| \quad \text{MHz} \tag{6-8}$$

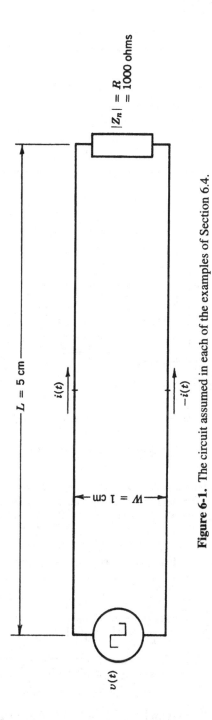

Figure 6-1. The circuit assumed in each of the examples of Section 6.4.

$L = 5$ cm

$W = 1$ cm

$i(t)$

$-i(t)$

$v(t)$

$|Z_n| = R$
$= 1000$ ohms

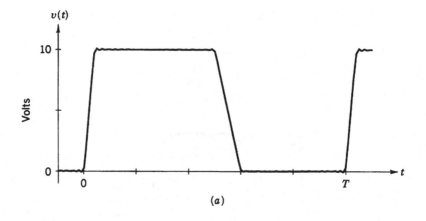

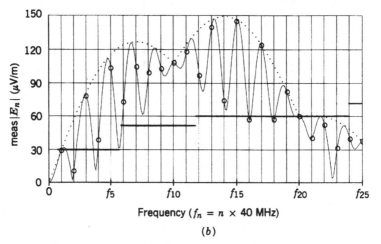

Figure 6-2. The initial source voltage waveshape in Example 6-1 and predicted values of measured radiated electric field strength: (*a*) voltage waveform with $t_R = T/25$, $t_F = T/10$, and $t_d = T/2$; (*b*) meas$|E_n|$ (○), Max(meas$|E_n|$) (···), and regulatory limits (—).

As a result, when this voltage waveform is the source voltage of the given circuit,

$$\text{Max(meas}|E_n|) = \frac{8000}{n\pi} \sin\left(\frac{11n\pi}{750}\right)$$

$$\cdot \left[\left|\sin\left(\frac{n\pi}{25}\right)\right| + 0.4\left|\sin\left(\frac{n\pi}{10}\right)\right|\right] \quad \mu\text{V/m} \qquad (6\text{-}9)$$

This equation gives the maximum possible value that meas$|E_n|$ can have for this circuit and source voltage at each measurement frequency f_n.

The values of meas$|E_n|$ caused by this circuit when this voltage waveform is the source, which are obtained from Eqs. 6-6 and 6-7, are shown in Fig. 6-2b. The values of the expression for meas$|E_n|$ are plotted as a continuous function of f, and the values that would be measured at each frequency f_n are circled. The values of Max(meas$|E_n|$), obtained with Eq. 6-9, are also plotted for comparison to meas$|E_n|$. Note that 16 values of meas$|E_n|$, which occur at frequencies from $f_3 = 120$ MHz to $f_{20} = 800$ MHz, equal or exceed the combined regulatory limits with this source voltage. Many of those values of meas$|E_n|$ are equal to more than twice the regulatory limit. To see how the waveshape of this voltage would have to be modified to make the circuit's radiations less than the regulatory limits, D_n will be modified using the following procedure.

Max(meas$|E_n|$) is plotted for several different voltage waveshapes to determine what changes in t_R and t_F will make this circuit pass regulatory limits at all frequencies. Those plots are shown in Fig. 6-3. It is seen in Fig. 6-3h that a source voltage in this circuit for which $t_R = T/5$ and $t_F = 2T/5$ will no longer cause Max(meas$|E_n|$) to exceed regulatory limits. That voltage is shown in Fig. 6-4a, and the passage of regulatory limits by meas$|E_n|$ is verified for each individual frequency f_n in Fig. 6-4b. Note that only a few values of meas$|E_n|$ attain the maximum at the frequencies f_n. Because of this, and because the rise and fall times of the voltage waveform of Fig. 6-4 are considerably longer than that of the original voltage, some other possible changes will be considered.

In Fig. 6-5, a voltage waveform with $t_R = 4T/25$, $t_F = 2T/5$, and $t_d = T/2$ is shown, together with the values of meas$|E_n|$ that will result if this is the circuit's source voltage. With this source voltage, the regulatory limits are met at all frequencies except $f_3 = 120$ MHz, where meas$|E_n|$ is perhaps 5 μV/m above the limit. Therefore, to illustrate one other step that might be taken, if this voltage were to be otherwise acceptable, t_d will be changed slightly. This will cause the individual values of meas$|E_n|$ to be changed, but it will not affect the maximum values, because the function Max(meas$|E_n|$) is independent of t_d.

The results of this trial-and-error investigation are shown in Fig. 6-6. It was found that by changing the value of t_d for this voltage waveform from $T/2$ to $4T/9$, all values of meas$|E_n|$ are moved below the regulatory limits. However, in practice, meas$|E_n|$ would probably not be safely below the limit at the frequencies f_3 and f_4. Nevertheless, a minor adjustment such as this to the source voltage waveshape may sometimes be useful in fully achieving control of troublesome circuit-current radiations.

Example 6-2   Now, suppose the same circuit were to have the source voltage shown in Fig. 6-7a with the same linear rise as that of the previous example, but with an exponential fall for which $t_F = 2T/5e$. With this source voltage waveform in the circuit, meas$|E_n|$ would have the values given in Fig. 6-7b.

To see what changes in this waveshape might be necessary to sufficiently reduce radiation, several plots of Max(meas$|E_n|$) are again made, as in Fig. 6-8. It appears that there is little hope of making meas$|E_n|$ acceptable by changing only the voltage waveshape by itself. Nevertheless, Fig. 6-8h indicates that

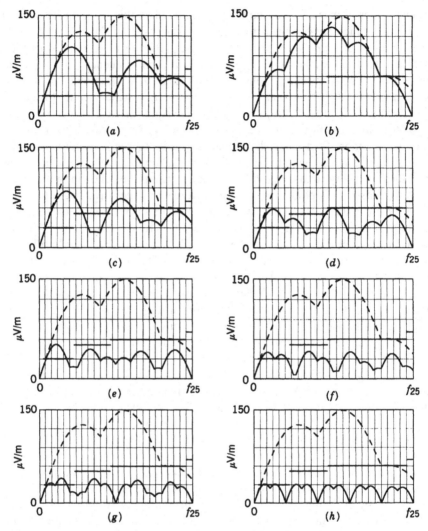

Figure 6-3. Max(meas$|E_n|$) for possible source voltage waveforms (—), and for the initial source voltage waveform (---) of Example 6-1: (a) $t_R = 2T/25$ and $t_F = T/10$; (b) $t_R = T/25$ and $t_F = T/5$; (c) $t_R = 3T/25$ and $t_F = T/10$; (d) $t_R = 3T/25$ and $t_F = T/5$ (e) $t_R = 4T/25$ and $t_F = T/5$; (f) $t_R = 4T/25$ and $t_F = 3T/10$; (g) $t_R = 4T/25$ and $t_F = 2T/5$; (h) $t_R = T/5$ and $t_F = 2T/5$.

with $t_R = T/5$ and $t_F = 3T/5e$, meas$|E_n|$ is liable to be slightly high in value only in the 30- to 230-MHz range. Therefore, voltage waveforms with those rise and fall times and different values of t_d will be briefly examined to see what might be accomplished by varying t_d. The first waveform considered is that of Fig. 6-9a, which has the rise and fall times $t_R = T/5$ and $t_F = 3T/5e$ as in Fig. 6-8h and for which $t_d = T/2$. It is seen in Fig. 6-9b that the only

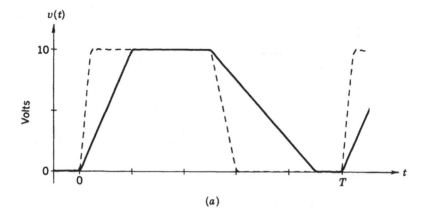

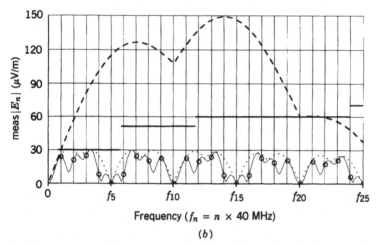

Figure 6-4. The first modified source voltage waveform considered in Example 6-1, and predicted values of measured radiated electric field strength: (*a*) voltage waveform with $t_R = T/5$, $t_F = 2T/5$, and $t_d = T/2$; (*b*) meas$|E_n|$ (○) and Max(meas$|E_n|$) (···) for this voltage waveform, Max(meas$|E_n|$) (---) for the initial voltage waveform, and regulatory limits (—).

frequency at which meas$|E_n|$ exceeds the limits is $f_3 = 120$ MHz. Therefore, it is possible that changing t_d may solve that problem.

A voltage waveform with the same rise and fall times and with $t_d = 2T/5$ provides a near solution to this problem, as shown in Fig. 6-10. However, this waveform is quite different from the original one of Fig. 6-7*a*, and while meas$|E_n|$ has been considerably reduced at the frequency f_3, it now exceeds the limit at f_4. Again, the implication of this example is that minor adjustments to a voltage waveshape may occasionally be useful, but they will seldom be sufficiently effective to fully control circuit-current radiations.

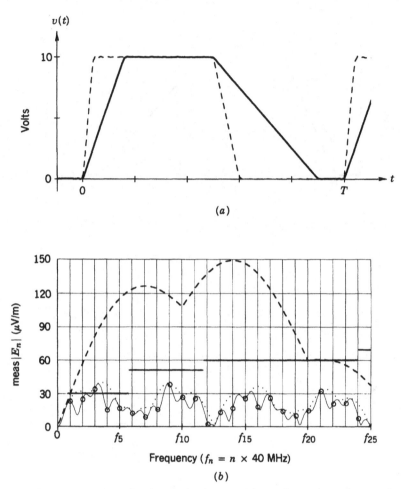

Figure 6-5. The second modified source voltage waveform considered in Example 6-1, and predicted values of measured radiated electric field strength: (*a*) voltage waveform with $t_R = 4T/25$, $t_F = 2T/5$, and $t_d = T/2$; (*b*) meas$|E_n|$ (○) and Max(meas$|E_n|$) (···) for this voltage waveform, Max(meas$|E_n|$) (---) for the initial voltage waveform, and regulatory limits (—).

Example 6-3   Next, suppose the same circuit has the source voltage that is illustrated in Fig. 6-11*a*. That voltage waveform has both an exponential rise and an exponential fall with $t_R = T/10e$, $t_F = 2T/5e$, and $t_d = T/2$. From Fig. 6-11*b* it is seen that this circuit and voltage do not even come close to satisfying the regulatory limits except at f_1, f_2, and f_6. To investigate further, Max(meas$|E_n|$) is again plotted for several alternative voltage waveforms in Fig. 6-12. It is quite clear from that figure that it will probably be impossible to reduce the radiation of this circuit and voltage satisfactorily by adjusting only the rise and fall times of this voltage waveform.

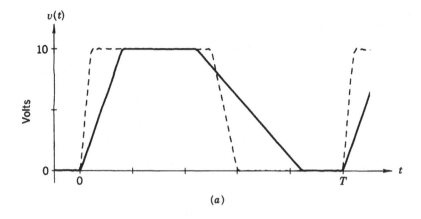

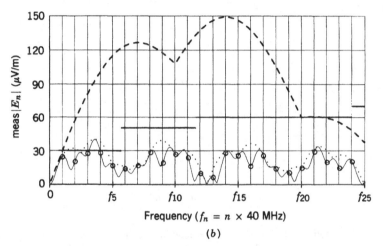

Figure 6-6. The third modified source voltage waveform considered in Example 6-1, and predicted values of measured radiated electric field strength: (a) voltage waveform with $t_R = 4T/25$, $t_F = 2T/5$, and $t_d = 4T/9$; (b) meas$|E_n|$ (○) and Max(meas$|E_n|$) (···) for this voltage waveform, Max(meas$|E_n|$) (---) for the initial voltage waveform, and regulatory limits (—).

It is also clear from Fig. 6-12 that plots of Max(meas$|E_n|$) take on quite a different look when both transitions are exponential. And in Fig. 6-11b, note also that for all of the frequencies $f_n = nf$ that are odd multiples of f, in every case, meas$|E_n| \cong$ Max(meas$|E_n|$). On the other hand, at every even multiple of f, meas$|E_n|$ is very nearly a minimum value. This happens when both transitions are exponential, because of the following relationships.

If both the rise and fall of $v(t)$ are exponential, then as f_n increases, the variables A_n, B_n, $\sin(A_n)$, and $\sin(B_n)$ all approach ($\rightarrow$) constants. As the frequency f_n gets large,

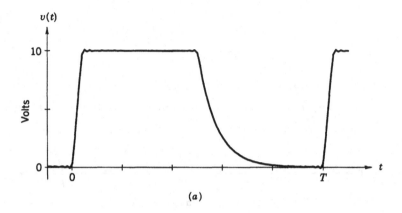

(a)

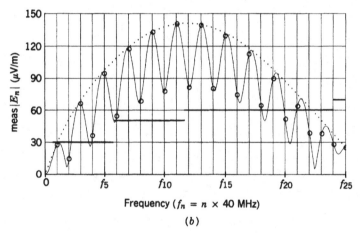

Frequency ($f_n = n \times 40$ MHz)

(b)

Figure 6-7. The initial source voltage waveshape in Example 6-2, and predicted values of measured radiated electric field strength: (a) voltage waveform with $t_R = T/25$, $t_F = 2T/5e$, and $t_d = T/2$; (b) meas$|E_n|$ (○), Max(meas$|E_n|$) (· · ·), and regulatory limits (—).

$$A_n = \arctan(\pi f_n t_R) \longrightarrow \arctan(+\infty) = \frac{\pi}{2}$$

$$B_n = \arctan(\pi f_n t_F) \longrightarrow \arctan(+\infty) = \frac{\pi}{2}$$

and

$$\sin(A_n) \longrightarrow \sin\left(\frac{\pi}{2}\right) = 1$$

$$\sin(B_n) \longrightarrow \sin\left(\frac{\pi}{2}\right) = 1$$

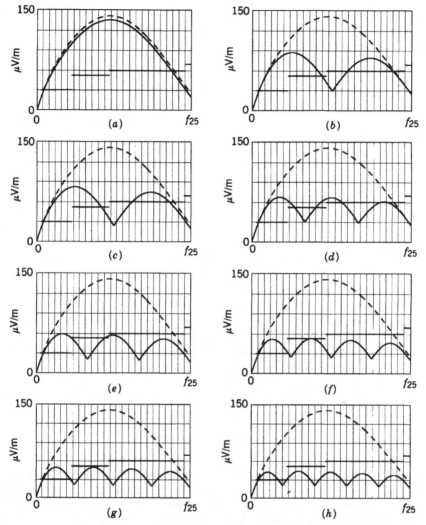

Figure 6-8. Max(meas$|E_n|$) for possible source voltage waveforms (—) and for the initial source voltage waveform (---) of Example 6-2: (*a*) $t_R = T/25$ and $t_F = T/2e$; (*b*) $t_R = 2T/25$ and $t_F = 2T/5e$; (*c*) $t_R = 2T/25$ and $t_F = T/2e$; (*d*) $t_R = 3T/25$ and $t_F = 2T/5e$; (*e*) $t_R = 3T/25$ and $t_F = 3T/5e$; (*f*) $t_R = 4T/25$ and $t_F = T/2e$; (*g*) $t_R = 4T/25$ and $t_F = 3T/5e$; (*h*) $t_R = T/5$ and $t_F = 3T/5e$.

Since $t_d = T/2$, and A_n and B_n are both approaching $\pi/2$, it also follows that

$$\cos(2\pi f_n t_d - A_n + B_n) = \cos(n\pi - A_n + B_n) \rightarrow \cos(n\pi) = \pm 1$$

As a result of all these things,

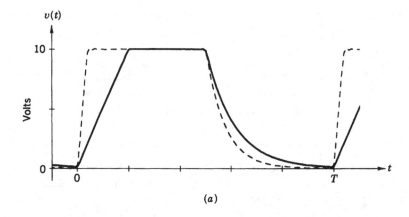

(a)

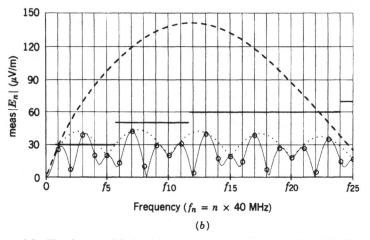

Frequency ($f_n = n \times 40$ MHz)

(b)

Figure 6-9. The first modified source voltage waveform considered in Example 6-2, and predicted values of measured radiated electric field strength: (*a*) voltage waveform with $t_R = T/5$, $t_F = 3T/5e$, and $t_d = T/2$; (*b*) meas$|E_n|$ ($\circ$), Max(meas$|E_n|$) with this voltage ($\cdots$) and with the original voltage (---), and regulatory limits (—).

$$D_n = \sqrt{\frac{\sin^2(A_n)}{t_R^2} - 2\,\frac{\sin(A_n)}{t_R}\,\frac{\sin(B_n)}{t_F}\,\cos(2\pi f_n t_d - A_n + B_n) + \frac{\sin^2(B_n)}{t_f^2}}$$

$$\longrightarrow \sqrt{\frac{1}{t_R^2} - 2\,\frac{1}{t_R}\,\frac{1}{t_F}\,\cos(n\pi) + \frac{1}{t_F^2}}$$

$$\longrightarrow \frac{1}{t_R} \mp \frac{1}{t_F} \tag{6-10}$$

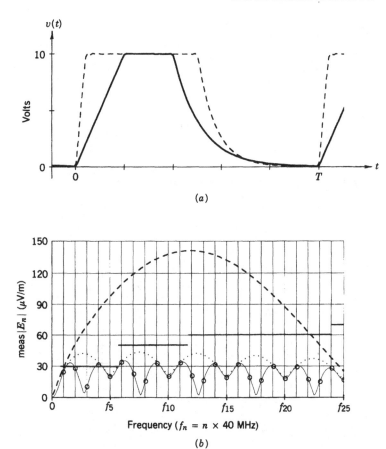

Figure 6-10. The second modified source voltage waveform considered in Example 6-2, and predicted values of measured radiated electric field strength: (*a*) voltage waveform with $t_R = T/5$, $t_F = 3T/5e$, and $t_d = 2T/5$; (*b*) meas$|E_n|$ (○), Max(meas$|E_n|$) with this voltage (···) and with the original voltage (---), and regulatory limits (—).

In other words, when both transitions of $v(t)$ are exponential and $t_d = T/2$, D_n approaches $1/t_R + 1/t_F$, for large odd values of n and $1/t_R - 1/t_F$, for large even values. For smaller values of n, for which $|\sin(A_n)| < 1$, and $|\sin(B_n)| < 1$, it follows that $D_n \cong |\sin(A_n)|/t_R \mp |\sin(B_n)|/t_F$.

In this example, then,

$$\text{meas}|E_n| \leq \frac{8 \times 10^{-6}}{n\pi} \sin\left(\frac{11n\pi}{750}\right)\left[\frac{|\sin(A_n)|}{t_R} + \frac{|\sin(B_n)|}{t_F}\right]$$

$$= \frac{8 \times 10^{-6}}{n\pi} \sin\left(\frac{11n\pi}{750}\right)\left[\frac{10e}{T}|\sin(A_n)| + \frac{5e}{2T}|\sin(B_n)|\right]$$

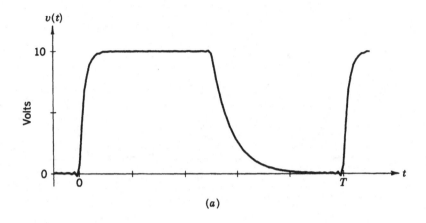

(a)

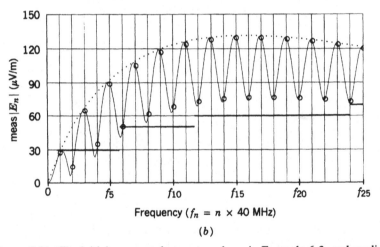

(b)

Figure 6-11. The initial source voltage wave shape in Example 6-3, and predicted values of measured radiated electric field strength: (a) Voltage waveform with $t_R = T/10e$, $t_F = 2T/5e$, and $t_d = T/2$; (b) meas$|E_n|$ (○), Max(meas$|E_n|$) (···), and regulatory limits (—).

$$= \frac{320 \times 5e}{n\pi} \sin\left(\frac{11n\pi}{750}\right) [2|\sin(A_n)| + 0.5|\sin(B_n)|]$$

$$= \sin\left(\frac{11n\pi}{750}\right) \left[\frac{3200e}{\sqrt{n^2\pi^2 + 100e^2}} + \frac{800e}{\sqrt{n^2\pi^2 + 6.25e^2}} \right] \quad \mu V/m$$

$$= \text{Max(meas}|E_n|) \tag{6-11}$$

And, when n is odd, meas$|E_n| \cong \text{Max(meas}|E_n|)$.

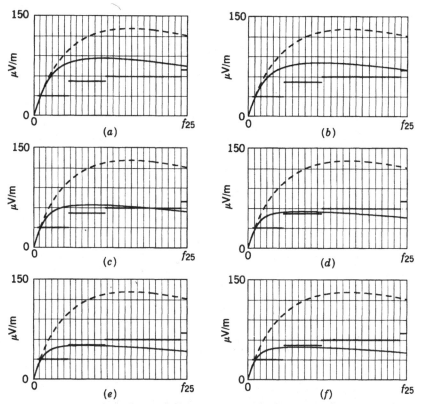

Figure 6-12. Max(meas$|E_n|$) for possible source voltage waveforms (—), and for the initial source voltage waveform (---) of Example 6-3: (*a*) $t_R = T/5e$ and $t_F = 2T/5e$; (*b*) $t_R = T/5e$ and $t_F = T/2e$; (*c*) $t_R = 3T/10e$ and $t_F = T/2e$; (*d*) $t_R = 2T/5e$ and $t_F = T/2e$; (*e*) $t_R = 2T/5e$ and $t_F = 3T/5e$; (*f*) $t_R = T/2e$ and $t_F = T/2e$.

Now, suppose that the voltage of Fig. 6-11*a* is replaced as the source voltage in the given circuit by the voltage of Fig. 6-13*a*, for which $t_R = t_F = T/2e$, as in Fig. 6-12*f*, and $t_d = T/2$. As seen in Fig. 6-13*b*, meas$|E_n|$ is well above the regulatory limits at both f_3 and f_5. It is perhaps of greater interest, however, that with this voltage as source voltage, meas$|E_n|$ and Max(meas$|E_n|$) are *equal* for every *odd* multiple of f, and meas$|E_n|$ = 0 for every *even* multiple of f. This happens because $t_R = t_F$ and $t_d = T/2$, because then $A_n = B_n$, and

$$\cos(\omega_n t_d - A_n + B_n) = \cos(2\pi n f t_d)$$
$$= \cos(n\pi)$$
$$= \pm 1 \qquad (6\text{-}12)$$

When n is even, $\cos(n\pi) = +1$, and when n is odd, $\cos(n\pi) = -1$. So, whenever n is even, $\cos(\omega_n t_d - A_n + B_n) = +1$, and it follows from Eq. 6-3 that

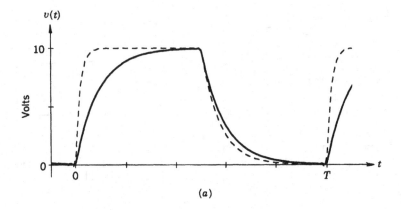

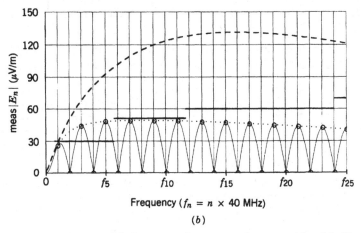

Figure 6-13. The first modified source voltage wave shape considered in Example 6-3, and predicted values of measured radiated electric field strength: (a) voltage waveform with $t_R = T/2e$, $t_F = T/2e$, and $t_d = T/2$; (b) meas$|E_n|$ (○), Max(meas$|E_n|$) with this voltage (···) and with the original voltage (---), and regulatory limits (—).

$$D_n = \sqrt{\left[\frac{\sin(A_n)}{t_R} - \frac{\sin(B_n)}{t_F}\right]^2}$$

$$= \left|\frac{\sin(A_n)}{t_R} - \frac{\sin(B_n)}{t_F}\right|$$

$$= 0 \qquad\qquad (6\text{-}13)$$

Therefore, if $t_R = t_F$, $t_d = T/2$, and n is even, then $D_n = 0$ and meas$|E_n| = 0$. Also, whenever n is odd, $\cos(\omega_n t_d - A_n + B_n) = -1$, and, since $\sin(A_n) = \sin(B_n)$, it follows from Eq. 6-3 that

$$D_n = \sqrt{\left[\frac{\sin(A_n)}{t_R} + \frac{\sin(B_n)}{t_F}\right]^2}$$

$$= \left|\frac{\sin(A_n)}{t_R} + \frac{\sin(B_n)}{t_F}\right|$$

$$= \frac{|\sin(A_n)|}{t_R} + \frac{|\sin(B_n)|}{t_F} \tag{6-14}$$

This is the maximum value of D_n for any frequency f_n. Thus, whenever n is odd, $t_d = T/2$ and $t_R = t_F$ and it is clear that meas$|E_n|$ = Max(meas$|E_n|$).

To see what the effect will be on meas$|E_n|$ if the rise and fall are both exponential, but $t_R \neq t_F$ and $t_d \neq T/2$, the source voltage of the given circuit will be changed to that of Fig. 6-14a. That voltage has an exponential rise time $t_R = 2T/5e$, an exponential fall time $t_F = 3T/5e$, and $t_d = 2T/5$. The effects these changes in the source voltage have on meas$|E_n|$ are shown in Fig. 6-14b. As might have been expected, meas$|E_n|$ is now below the regulatory limit at the frequencies f_3 and f_5. However, now the limit is exceeded at f_4 by about 15 μV/m, whereas, with the previous source voltage wave shape, meas$|E_n|$ was zero at the frequency f_4.

For one last experiment, suppose the source voltage waveshape of Fig. 6-14 is replaced with that of the voltage shown in Fig. 6-15, for which $t_R = 3T/5e$, $t_F = 2T/5e$, and $t_d = t_r = 3T/5$. The combined regulatory limits are still exceeded by meas$|E_n|$ at f_4, by the same amount as before. In fact, although the continuous graphs of meas$|E_n|$ differ considerably in Figs. 6-14 and 6-15, the measured values in those figures are equal at every frequency f_n.

There is a fairly obvious reason for meas$|E_n|$ having the same values at each frequency f_n for these two voltages, and it can be seen as follows. Let $v(t)$ be the voltage waveform of Fig. 6-14a, and let $v'(t)$ be the voltage waveform of Fig. 6-15a. Then, because the rise and fall times of $v(t)$ and $v'(t)$ have the same values, but they are interchanged, and because $t'_d = T - t_d$, it follows that $v'(t) = V_p - v(t - t_d)$. Or, in the frequency-domain,

$$v(t) = \overline{v(t)} + \sum_{n=1}^{\infty} |V_n| \cos(\omega_n t - \phi_n)$$

and

$$v'(t) = [V_p - \overline{v(t)}] - \sum_{n=1}^{\infty} |V_n| \cos[\omega_n(t - t_d) - \phi_n] \tag{6-15}$$

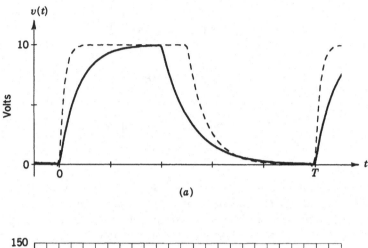

(a)

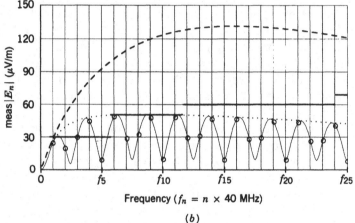

(b)

Figure 6-14. The second modified source voltage wave shape considered in Example 6-3, and predicted values of measured radiated electric field strength: (a) voltage waveform with $t_R = 2T/5e$, $t_F = 3T/5e$, and $t_d = 2T/5$; (b) meas$|E_n|$ (○) and Max(meas$|E_n|$) with this voltage (···) and with the original voltage (---) and regulatory limits (—).

Thus, at every frequency f_n the amplitudes of the sinusoidal components of $v(t)$ and $v'(t)$ have the same magnitude $|V_n|$. As noted in the previous chapter, the amplitude of the nth sinusoidal current component is $|I_n| = |V_n|/|Z_n|$, and meas$|E_n|$ is proportional to $|I_n|$. Therefore, if either $v(t)$ or $v'(t)$ is the source voltage for the same circuit, meas$|E_n|$ will be the same for either at each measurement frequency f_n. This will be the case, despite the fact that continuous plots of meas$|E_n|$ for $v(t)$ and $v'(t)$ will be very different.

It should be clear from this example, if not from earlier examples, that if frequent adjustments to D_n are to be part of controlling circuit-current radiations, a computer should be close at hand. The transition factor is a well-defined

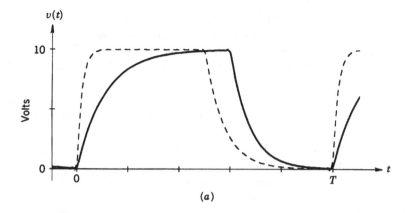

(a)

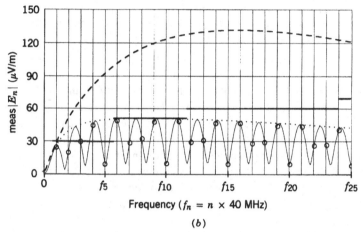

(b)

Figure 6-15. The third modified source voltage wave shape considered in Example 6-3, and predicted values of measured radiated electric field strength: (*a*) voltage waveform with $t_R = 3T/5e$, $t_F = 2T/5e$, and $t_d = 3T/5$; (*b*) meas$|E_n|$ (○), Max(meas$|E_n|$) with this voltage (···) and with the original voltage (---), and regulatory limits (—).

mathematical function, but its variations in value from one frequency f_n to the next make meas$|E_n|$ not easily estimated. If a change in voltage waveshape is being considered, values of meas$|E_n|$ should be computed at each frequency f_n for each waveshape and placed in a table or graph for comparison at each frequency f_n.

6.4.1 Equivalent Transitions

A characteristic of linear and exponential voltage transitions that can sometimes be useful in reducing and controlling circuit-current radiations is defined here as transitional equivalence. In many cases a voltage transition need not be 100%

complete to cause the desired response in a circuit. Many circuits will respond to voltage transitions well before the transitions are completed, and whether the voltage change occurs linearly or exponentially makes no difference. Thus, if a circuit responds the same when either a linear or an exponential voltage transition reaches a certain level and that level is reached in the same time, the transitions are equivalent for that circuit. *Equivalent transitions* reach the same percentage of change in the same amount of time and cause the same circuit response.

For example, if linear and exponential transitions both complete 80% of their transition in the same amount of time, they will be said to be 80% equivalent. For voltage waveforms with rises that are 80% equivalent, the time required for either to go from 0 to $0.8V_p$ is $0.8t_R$, where $t_R = t_r$ for linear rises and $t_R = t_r/e$ for exponential rises. For voltage waveforms with falls that are 80% equivalent, the time required for either to go from V_p to $0.2V_p$ is $0.8t_F$, where $t_F = t_f$ for linear falls and $t_F = t_f/e$ for exponential falls. This is illustrated graphically in Fig. 6-16, together with examples of 90 and 95% equivalence. The notation t_{RE} is used to distinguish $t_R = t_r/e$ for exponential rises from $t_R = t_r$ for linear rises, and t_{FE} is used to distinguish $t_F = t_f/e$ for exponential falls from $t_F = t_f$ for linear falls.

Thus, it is seen in Fig. 6-16 that, when rises are 80% equivalent, $t_R = t_{RE}$, and when falls are 80% equivalent, $t_F = t_{FE}$. However, if rises are 90% equivalent, $t_R = 1.28t_{RE}$, and if falls are 90% equivalent, then $t_F = 1.28t_{FE}$. Also, 95% equivalence implies that $t_R = 1.58t_{RE}$ for rises and $t_F = 1.58t_{FE}$ for falls. In other words, the ratios of linear-to-exponential transition times, t_R/t_{RE} or t_F/t_{FE}, are 1.0 for 80% equivalence, 1.28 for 90% equivalence, and 1.58 for 95% equivalence. Other ratios for different percentages of equivalence are given in Fig. 6-17. The equations with which that graph is plotted are

$$\frac{t_R}{t_{RE}} = \frac{t_F}{t_{FE}} = \frac{50}{p} \ln\left(\frac{100}{100-p}\right) \qquad (6\text{-}16)$$

where p is the specified percentage for equivalence.

The implication of these relationships is rather obvious. If equivalent voltage transitions cause a circuit to operate properly, and equivalence is based on the transitions being anywhere from 80 to 100% complete, less radiation will be caused by linear transitions. The reason for this is that the linear transitions will have greater values of t_R and t_F, causing D_n to be less for them than it is for their exponential equivalents. Some of the effects of transitional equivalence are illustrated in the following example.

Example 6-4 Consider the exponentially rising and falling voltage waveform of Fig. 6-18a. The rise time of that waveform is $t_R = t_r/e = T/25$ and its fall time is $t_F = t_f/e = T/10$. Therefore, this voltage waveform and the linearly rising and falling voltage waveform of Fig. 6-2 are 80% equivalent. Graphs of

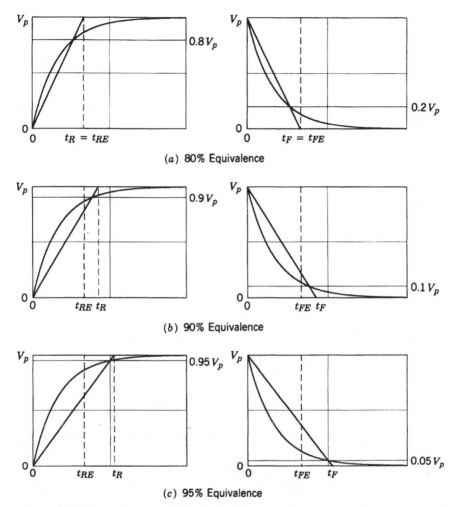

Figure 6-16. Transitional equivalence based on equal percentages of completion of rises or falls in equal periods of time.

Max(meas$|E_n|$) when each of these voltage waveforms is the source voltage for the circuit of Fig. 6-1 are shown in Fig. 6-19a. Also shown in Fig. 6-19 are graphs of Max(meas$|E_n|$) for several other linearly rising and falling source voltage waveforms that are equivalent to the waveform of Fig. 6-18 for the percentages indicated. These voltage waveforms and the values of meas$|E_n|$ that result when each is the source voltage of the circuit of Fig. 6-1 are illustrated in Figs. 6-20 and 6-21.

From all of the above examples, it appears that adjustment of the transition factor D_n will generally involve trial and error, despite the fact that it is a well-defined mathematical function. Nevertheless, a few general observations about D_n and its role in minimizing meas$|E_n|$ can be made based on the preceding

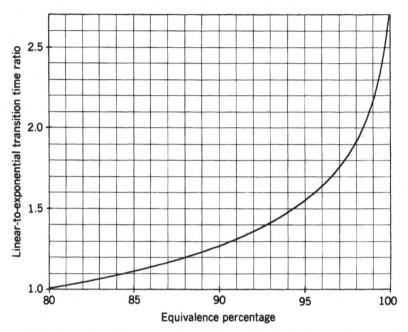

Figure 6-17. The ratio of linear-to-exponential transition times of equivalent transitions as a function of the percentage of equivalence.

examples:

1. Generally, the rise time t_R and the fall time t_F should be no shorter than absolutely necessary.
2. If $t_d = T/2$ and both transitions are exponential, meas$|E_n|$ will be at or very near its maximum at every frequency f_n that is an odd multiple of f.
3. The value of t_d can sometimes be varied slightly to reduce meas$|E_n|$ at one or perhaps a few troublesome frequencies f_n.
4. Linear voltage transitions cause less radiation than equivalent exponential transitions whenever equivalence occurs when the transitions are more than 80% complete.

6.5 CIRCUIT-CURRENT PATH WIDTH

As previously noted, the width of a circuit current's path is of primary importance in controlling the radiations of the current. The circuit or path width W is the distance between the paths of the currents that connect the source to the loads of a circuit. For all of the rectangular circuits discussed up to this point—large, medium, or small—meas$|E_n|$ is directly proportional to W, as

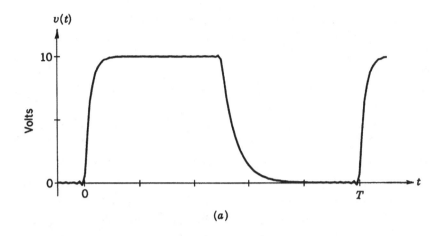

(a)

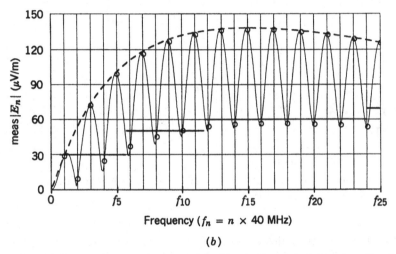

Frequency ($f_n = n \times 40$ MHz)

(b)

Figure 6-18. The initial source voltage waveform in Example 6-4, and predicted values of measured radiated electric field strength: (a) voltage waveform with $t_R = T/25$, $t_F = T/10$, and $t_d = T/2$; (b) meas$|E_n|$ (o), Max(meas$|E_n|$) (---), and regulatory limits (—).

summarized in Eq. 6-1. However, in practice, most circuits are not perfectly rectangular like those that have been considered so far. Nevertheless, as mentioned in Chapter 1, one good way to attack and solve any problem is to break it down into simpler parts. A good way to obtain a better understanding of a relatively complicated problem is to first consider simpler problems that are similar to it. Therefore, in this section, based on previous results and observations, a general approach is laid out for effectively minimizing the radiations of circuits of any given shape.

Before going on, however, one example of the effectiveness of reducing W will be given that is partially related to the discussions of the previous section.

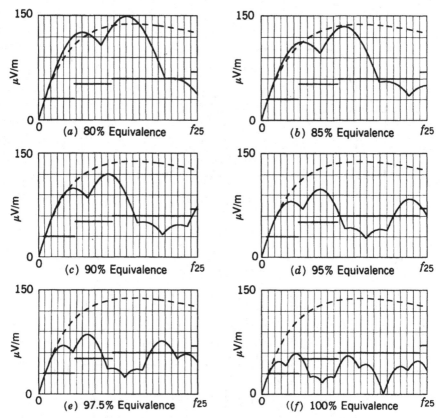

Figure 6-19. Max(meas$|E_n|$) for the initial source voltage of Example 6-4 (---) compared to several of its linear equivalents (—).

Example 6-5 Consider the circuit of Fig. 6-1 when it has a source voltage with the linearly rising and falling waveshape of Fig. 6-2a. It was seen in Example 6-1 that to sufficiently reduce the radiation from that circuit by changing only the source voltage, its wave shape would have to be changed considerably. As seen in Fig. 6-2b, with that voltage, at the frequency f_5, meas$|E_n|$ is almost 3.5 times the regulatory limit, and at f_{15} it is nearly 2.5 times the limit. However, at all other measurement frequencies, meas$|E_n|$ is not so high relative to the limits as it is at those two frequencies.

Thus, for W to be reduced sufficiently for this circuit and voltage to pass at all frequencies, including f_5, it would have to be reduced by a factor of $\frac{1}{3.5} \cong 0.29$ or less. However, since meas$|E_n|$ is less than 2.5 times the regulatory limit at all other frequencies, a reduction factor of 0.4 would probably suffice at those frequencies. The results of reducing the width of the circuit of Fig. 6-1 from 1 to 0.4 cm are shown in Fig. 6-22. The voltage is unchanged and at the frequency f_3, meas$|E_n|$ equals the limit of 30 μV/m, and at f_5, it equals about 40 μV/m, which is still 10 μV/m above the limit. At all the other measurement

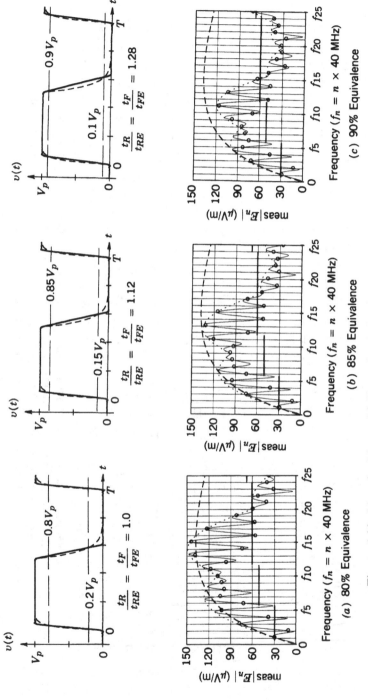

Figure 6-20. Linearly rising and falling voltage waveforms (---) that are equivalent to the initial voltage waveform (---) of Example 6-4, meas$|E_n|$ (o) and Max(meas$|E_n|$) (···) for the equivalent voltage waveforms, and Max(meas$|E_n|$) (---) for the initial voltage waveform, when any one of them is the source voltage in the circuit of Fig. 6-1.

169

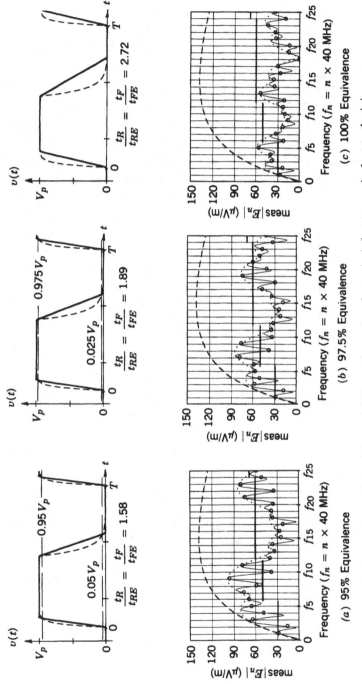

Figure 6-21. Linearly rising and falling voltage waveforms (—) that are equivalent to the initial voltage waveform (---) of Example 6-4, meas$|E_n|$ (o) and Max(meas$|E_n|$) (···) for the equivalent voltage waveforms, and Max(meas $|E_n|$) (---) for the initial voltage waveform, when any one of them is the source voltage in the circuit of Fig. 6-1.

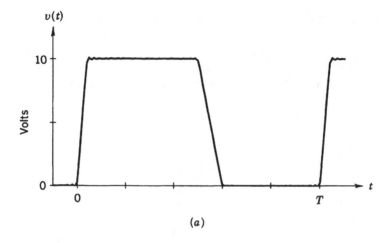

(a)

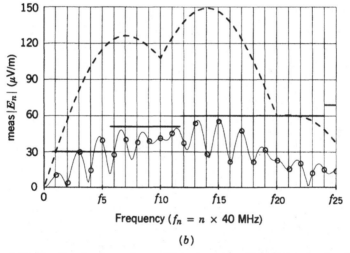

Frequency ($f_n = n \times 40$ MHz)

(b)

Figure 6-22. Predicted values of meas$|E_n|$ when the width of the circuit of Fig. 6-1 is reduced from 1.0 to 0.4 cm, and the source voltage is that of Example 6-1: (a) voltage waveform with $t_R = T/25$, $t_F = T/10$, and $t_d = 0.5T$; (b) meas$|E_n|$ (○) of the modified circuit, Max(meas$|E_n|$) (- - -) of the original circuit, and regulatory limits (—).

frequencies, however, meas$|E_n|$ is well below the regulatory limits with only this reduction in circuit width. Therefore, as observed in the previous section, a slight change in the time separating the voltage's transitions t_d may be in order here.

A look at Fig. 6-22b shows that a slight increase in the rate at which the continuous plot of meas$|E_n|$ varies up and down might bring its values at f_3 and f_5 below the limit. Thus, a slight increase in t_d may solve the problem, because that will increase the rate at which $\cos(\omega_n t_d - A_n + B_n)$ and meas$|E_n|$ vary. Indeed, as shown in Fig. 6-23b, it is seen after a few trials that an increase in t_d from $0.5T$ to $0.53T$ brings meas$|E_n|$ well within the limit at all frequencies.

This example illustrates the power of reductions in circuit width and the usefulness of changing t_d when meas$|E_n|$ is a problem at only one or two fre-

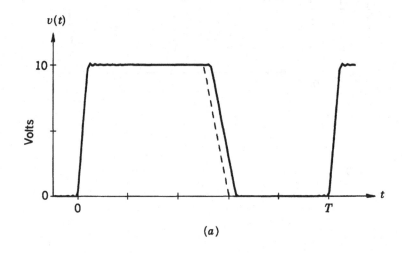

(a)

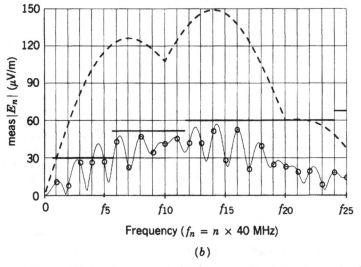

Frequency ($f_n = n \times$ 40 MHz)

(b)

Figure 6-23. Predicted values of meas$|E_n|$ when the width of the circuit of Fig. 6-1 is reduced from 1.0 to 0.4 cm, and t_d of the source voltage is increased as shown above: (a) Voltage waveform with $t_R = T/25$, $t_F = T/10$, and with t_d increased from $0.5T$ (---) to $0.53T$ (—); (b) meas$|E_n|$ (○) of the modified circuit and source voltage, Max(meas$|E_n|$) (---) of the original circuit and voltage, and regulatory limits (—).

quencies. As seen in Fig. 6-23a, the modified voltage wave shape is little different than the original, and a reduction of the circuit width from 1 to 0.4 cm is otherwise sufficient to solve the problem.

6.5.1 Other Circuit Shapes

Suppose the source and load of the circuit of Fig. 6-1 are such that the circuit's width cannot be further reduced in the immediate vicinity of the source and load. However, a short distance away from both the source and load, the width of the current's path can be reduced. The effect of such a reduction in the width of a current's path on $\text{meas}|E_n|$ can be closely approximated as follows.

Suppose the original dimensions of the circuit are those of the circuit of Fig. 6-1—$L = 5$ cm and $W = 1$ cm. Then at $f = 1000$ MHz, $W_\lambda = \frac{1}{30}$ and $L_\lambda = \frac{1}{6}$, so that at all frequencies of concern the circuit is either small or narrow and of medium length. Therefore, the expression describing its $\text{meas}|E_n|$ will be that of Eq. 6-1a. The shape and dimensions of the modified circuit are illustrated in Fig. 6-24c. From that figure, the average width of the modified circuit can be seen to be

$$\overline{W} = \frac{kW(hL) + (1 - h)LW}{L}$$

$$= [h(k - 1) + 1]W \tag{6-17}$$

And the maximum radiated electric field of the modified circuit will be approximately equal to a rectangular circuit with the same current, the length L, and the width $\overline{W}$. Therefore, from Eqs. 6-1 and 6-17, the modified circuit will be such that

$$\text{meas}|E_n| \cong \left(\frac{2Z_0}{n\pi^2 cd} \right) \frac{V_p D_n}{|Z_n|} \overline{W} \sin\left[\frac{n\pi f(2L + \overline{W})}{c} \right] \tag{6-18}$$

Several examples of the effects on $\text{meas}|E_n|$ of this partial reduction of the path width of a rectangular circuit are given for the original source voltages of Examples 6-1, 6-2, and 6-3, in Figs. 6-24, 6-25, and 6-26.

6.5.2 Connecting-Current Symmetry

Although, it has not been emphasized up to this point, the negative symmetry of the connecting currents of a circuit is extremely important in reducing the radiations of those currents. It was assumed for each of the circuits discussed so far that the source of the circuit's current was located at the center of one of the current elements that define the width of the rectangular circuit. That made the source equidistant to adjacent points along the paths of the currents

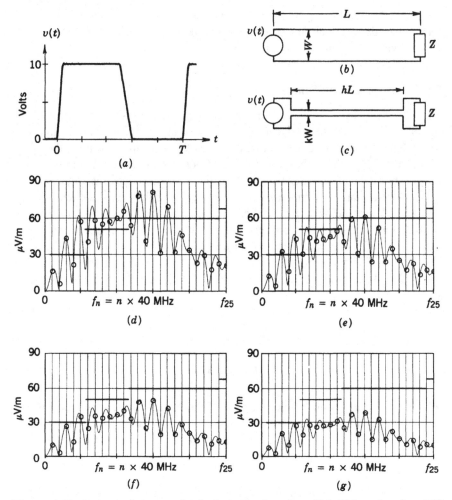

Figure 6-24. Reductions in meas$|E_n|$ obtained when the width W is reduced to kW over the length hL, and the source voltage is that of Example 6-1: (a) source voltage; (b) initial circuit; (c) modified circuit; (d) meas$|E_n|$ when $h = 0.6$ and $k = 0.3$; (e) meas$|E_n|$, when $h = 0.7$ and $k = 0.2$; (f) meas$|E_n|$, when $h = 0.8$ and $k = 0.2$; (g) meas$|E_n|$, when $h = 0.9$ and $k = 0.1$.

connecting the source and load. As a result, the currents that propagated along the segments from the source to the load were always equal in amplitude and oppositely directed at all adjacent points along those segments. That is precisely why moving the segments closer together—increasing their adjacency—causes greater cancellation in their radiations.

In moving the paths of a circuit's connecting currents closer together, it is extremely important that this symmetry be preserved if it exists, or that it be obtained if it does not exist. Consider the circuit of Fig. 6-27a. It can be seen

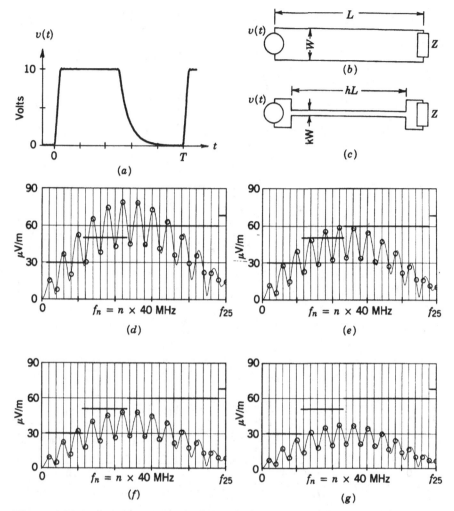

Figure 6-25. Reductions in meas$|E_n|$ obtained when the width W is reduced to kW over the length hL, and the source voltage is that of Example 6-2: (a) source voltage; (b) initial circuit; (c) modified circuit; (d) meas$|E_n|$, when $h = 0.6$ and $k = 0.3$; (e) meas$|E_n|$, when $h = 0.7$ and $k = 0.2$; (f) meas$|E_n|$, when $h = 0.8$ and $k = 0.2$; (g) meas$|E_n|$, when $h = 0.9$ and $k = 0.1$.

that connecting-current symmetry exists, because the phase constants of the currents are the same at each pair of points equidistant from the source, and the signs of the currents are opposite. However, W is obviously far from minimal. Thus, if there are no physical impediments in the actual implementation of the circuit, W should clearly be reduced. Because there are multiple loads, however, care must be taken in rerouting the currents. As shown in Fig. 6-27b, the current paths can be lengthened and moved closer together with the desired

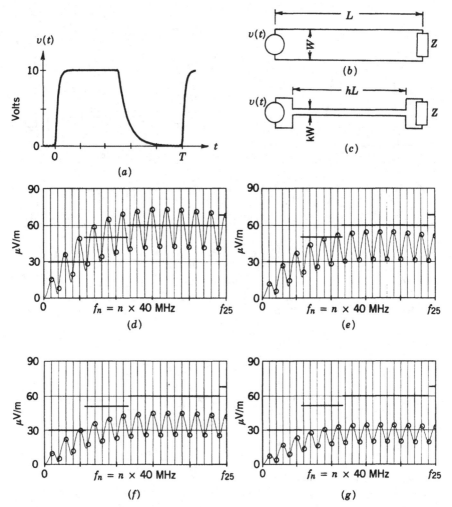

Figure 6-26. Reductions in meas$|E_n|$ obtained when the width W is reduced to kW over the length hL, and the source voltage is that of Example 6-3: (a) source voltage; (b) initial circuit; (c) modified circuit; (d) meas$|E_n|$, when $h = 0.6$ and $k = 0.3$; (e) meas$|E_n|$, when $h = 0.7$ and $k = 0.2$; (f) meas$|E_n|$, when $h = 0.8$ and $k = 0.2$; (g) meas$|E_n|$, when $h = 0.9$ and $k = 0.1$.

negative symmetry everywhere except for a short distance at each corner where the currents change direction.

The halfway point indicated in each of the circuits of Fig. 6-27 is the point equidistant from the source at which the propagating forward and return currents are one and the same. If a circuit's paths are properly laid out, that point, like the center of the source, will also be a center of symmetry. In other words, equal distances from that point along the forward and return current paths, the phases

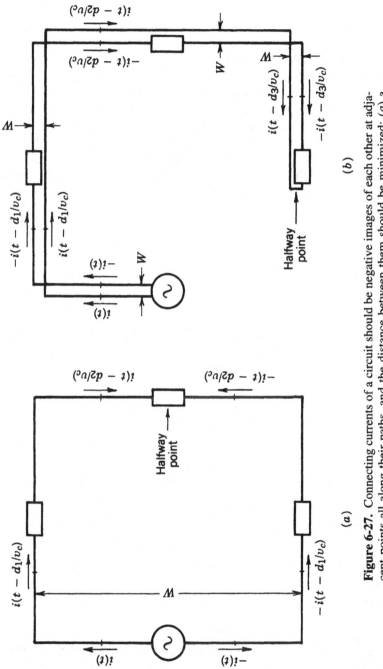

Figure 6-27. Connecting currents of a circuit should be negative images of each other at adjacent points all along their paths, and the distance between them should be minimized: (*a*) a circuit with symmetrical connecting currents, but *W* not minimized; (*b*) the same circuit with connecting current symmetry maximized and *W* minimized.

of the currents will be such that they are always equal in amplitude and oppositely directed.

Negative symmetry and closeness of the connecting currents of a circuit can sometimes be approximated in practice, based on the principle illustrated in Fig. 6-28. That principle says that current segments positioned above an infinite conducting plane, as illustrated in Fig. 6-28a, will establish negative images of themselves in the plane. The images are created by the radiated electric and magnetic fields of the currents that are induced on the surface of the plane by the current segments above it. As the current segments are brought closer and closer to the surface of the plane, the total radiated fields everywhere above the plane will be brought to zero. Thus, the radiated fields above the infinite conducting plane are the same as those that would be caused by the configuration illustrated in Fig. 6-28b. There the induced surface currents have been replaced by the

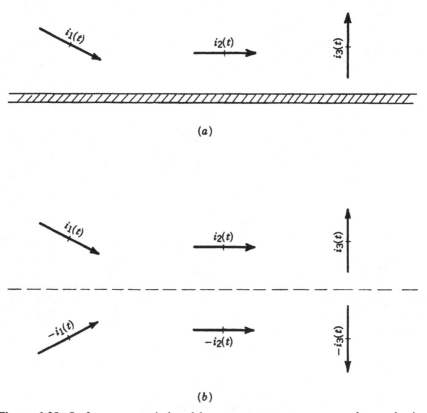

(a)

(b)

Figure 6-28. Surface currents induced by current segments on a nearby conducting plane establish the same radiated fields above the plane that would be established by negative images of the current segments with the conducting plane removed: (a) current segments above an infinite conducting plane; (b) the same current segments and their negative images relative to the previous position of the infinite conducting plane.

negative images of the current segments, and the plane has been removed. The method of analyzing the radiations of currents over a conductive plane, which is based on this equivalence, is known as the method of images.

Finite conductive ground planes are quite often used in practice to carry return connecting currents, on printed circuit boards, for example. This allows W to be made very small. However, whenever this is done, it should be kept in mind that (1) the ground plane is not infinite as assumed in Fig. 6-28, and (2) negative symmetry must be preserved if the currents' radiations are to cancel effectively. Because the ground plane is not infinite, the currents induced on it will not be exact images of the inducing currents, and cancellation will be reduced. For that reason, circuits with the greatest tendency to radiate should always be located near the center of a ground plane, and that ground plane should be no smaller than absolutely necessary. If negative symmetry is not preserved, the source will not be equidistant from adjacent points of the con- necting currents' paths, the currents will not be equal and opposite at those points, and cancellation will be further reduced.

One possible way to preserve negative symmetry when a ground plane is used is illustrated in Fig. 6-29. The center of the source is positioned flush with the ground plane. However, this implies that a hole must be made in the ground plane, and, depending on the size of the hole that is necessary, that will disturb the surface currents that are needed to establish symmetry in the first place.

One other point should be made about ground planes, their size, and connecting-current symmetry. As frequency increases, the size of a ground

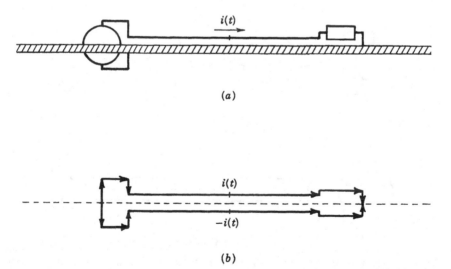

(a)

(b)

Figure 6-29. Circuit width can be effectively reduced with a finite ground plane, if neg- ative symmetry of the connecting currents can be obtained: (a) circuit implementation to obtain connecting-current symmetry; (b) desired symmetrical current pattern.

plane effectively becomes larger, and that is good. On the other hand, any lack of negative symmetry that exists at low frequencies becomes worse as frequency increases, because the wavelength becomes shorter. Thus, it should be clear that finite ground planes have their limitations, so far as the reduction and control of circuit-current radiations are concerned. Any attempts to minimize meas$|E_n|$ with ground planes should be very carefully thought out, and their effectiveness should always be verified experimentally.

Other ways to reduce W are illustrated in Fig. 6-30. The radiation of the

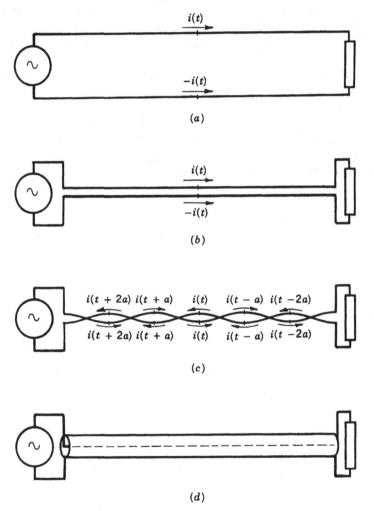

Figure 6-30. Progressively more effective methods of reducing connecting current radiations: (*a*) unmodified circuit; (*b*) closely parallel connecting currents; (*c*) closely twisted connecting currents; (*d*) coaxial connecting currents.

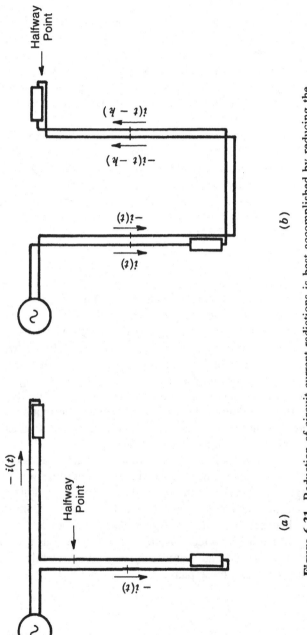

Figure 6-31. Reduction of circuit-current radiations is best accomplished by reducing the spacing between connecting currents and maintaining their negative symmetry: (*a*) no symmetry; (*b*) maximum symmetry.

circuit of Fig. 6-30a can be greatly reduced by simply making the connecting current paths between the source and load parallel and as close together as possible, as shown in Fig. 6-30b. If those results are still inadequate, then the parallel pair can be converted to a twisted pair, as illustrated in Fig. 6-30c. Greater reduction is accomplished with the twisted pair, because from any observation point, the twisted wires appear to be a series of small circuits with currents in adjacent circuits that are equal and opposite. Therefore, the residual radiations of each pair of adjacent twists will tend to further cancel each other.

Last, and best, of these approaches, of course, is the coaxial cable. The effect of a coaxial cable is to reduce W to zero for the length of the cable. Thus, if symmetry is preserved, the currents in the cable are equal in magnitude, oppositely directed, and from any external observation point they appear to follow exactly the same path. Therefore, over the length of the cable, the radiations of the currents will cancel completely.

6.6 SUMMARY

It has been seen here that the ultimate solution to minimizing circuit-current radiations is to minimize the distance W between the connecting currents of a circuit, while at the same time preserving their negative symmetry. Negative symmetry requires that adjacent points along connecting current paths be equidistant to the center of the circuit's source. The difference between connecting-current symmetry and the lack of it is illustrated in Fig. 6-31.

Other methods of reducing circuit current radiations that were discussed in this chapter are not so widely applicable as modifying a current's path. However, they do provide means of making additional adjustments when narrowing of the path is physically limited for one reason or another. Those methods include maximizing a circuit's impedance to reduce the amplitudes of radiating currents as much as possible. They also include changing the wave shape of the source voltage of a circuit by minimizing its amplitude, maximizing its rise and fall times, choosing the kinds of rises and falls—linear or exponential—and varying the time between rise and fall. Again, these methods are occasionally useful, but they are neither so widely useful nor so effective as narrowing the distance between the forward and return paths of a current, while maintaining its negative symmetry along those paths.

CHAPTER 7

CONTAINMENT OF UNINTENTIONAL RADIATIONS

7.1 INTRODUCTION

Containment of unintentional electromagnetic radiation is based on principles that have already been discussed, but they will now be differently applied. For example, a conducting plane in which a nearby current establishes its negative image can simply be viewed as a reflector of the radiation of that current. Based on that viewpoint then, a radiation that is fully reflected back toward its source from all directions will be fully contained. Previously it was seen that radiations can be effectively controlled by properly placing a reflector so that incident and reflected radiations cancel each other. In implementing containment, however, radiated fields are simply reflected back in the direction of their source. There is generally no attempt to cause any cancellation with the reflections.

Thus, the principles on which control and containment are based are the same and, with respect to the surrounding environment, the desired end results are the same. However, the methods of implementation are quite different from one another, and the implications of those differences are well worth examining.

7.2 FULLY CONTAINED RADIATING CURRENTS

The idea of containing circuit-current radiations is based primarily on the principle that electromagnetic fields do not exist in the interior of a perfect conductor. As discussed in Chapter 1, that makes the tangential electric field and the normal magnetic field both zero at the surface of a perfect conductor. Therefore, a radiating current placed inside a fully closed, perfectly conducting container can cause no tangential electric field, and no normal magnetic field, at the outer surface(s) of its container, or beyond.

Consider, for example, an element of current $i(t) = |I| \sin(\omega t)$ centered along the z axis of the coordinate system. The radiated electric field in the θ direction, which is everywhere tangential to the surface of an imaginary sphere of radius d in which the current element is centered, would be

$$E_\theta(t) = E_e(\theta) \cos\left[\frac{\omega(t-d)}{c}\right]$$

where

$$E_e(\theta) = \frac{Z_0 |I| L_{e\lambda}}{2d} \sin\theta \qquad (7\text{-}1)$$

However, as shown in Fig. 7-1, if the current element were also centered in a smaller, perfectly conducting spherical container, then on the sphere of radius d the tangential electric field is $E_\theta(t) = 0$.

On the other hand, the normal electric field and the tangential magnetic field at the outer surface of a perfectly conducting container of a radiating current will not necessarily be zero. Those field components are independent of the nonexistence of fields in the interior of a perfect conductor. As previously discussed in Chapter 1, those field components are determined by the charge and the current on the conductor's surfaces. Any charge or current induced on the inner surface of a perfectly conducting container will change the charge and possibly cause a current on the outer surface. In other words, if the total charge on a container is zero, the instantaneous charges on its inner and outer surfaces will always be equal and opposite, so that their sum is zero. However, the charge on either surface will generally change when radiation is incident on the other, and a nonzero surface charge will cause a nonzero electric field normal to the surface. The implications of this for a sinusoidal current element that is fully contained by a perfectly conducting sphere, in comparison to when it is not contained, are illustrated in Fig. 7-2.

Thus, it can be seen from Figs. 7-1 and 7-2 that a fully closed, perfectly conducting, spherical container would change the direction of the radiated electric field of a current element. Also, because there are no fields between the surfaces of a perfect conductor, in otherwise field-free space the time-varying charge on the outer surface would always be evenly distributed. Therefore, the magnitude of $E_r(t)$, the radially directed electric field in Fig. 7-2b, would be equal in all directions. In some directions $E_r(t)$ would be reduced compared to the original field $E_\theta(t)$, and in other directions it would be increased.

To reduce the radiation caused by surface charges induced on the outer surface of a perfectly conducting container of a radiating current, the surface can be connected to ground. However, while this will return the outer surface charge to

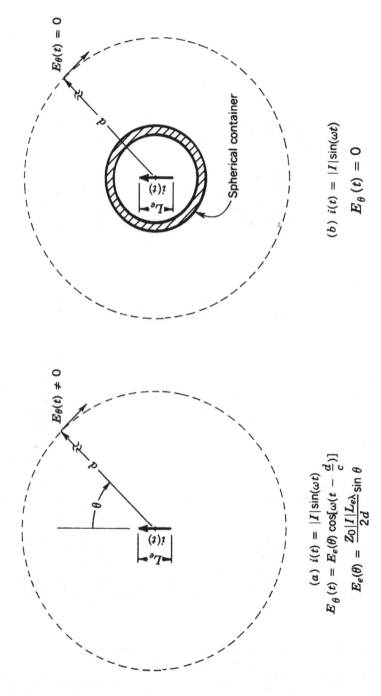

(a) $i(t) = |I|\sin(\omega t)$

$E_\theta(t) = E_e(\theta)\cos[\omega(t - \frac{d}{c})]$

$E_e(\theta) = \frac{Z_0|I|L_e\lambda}{2d}\sin\theta$

(b) $i(t) = |I|\sin(\omega t)$

$E_\theta(t) = 0$

Figure 7-1. The electric fields $E_\theta(t)$ from a sinusoidal current element (a) in free space, and (b) in a perfectly conducting, completely closed, spherical container, in otherwise free space.

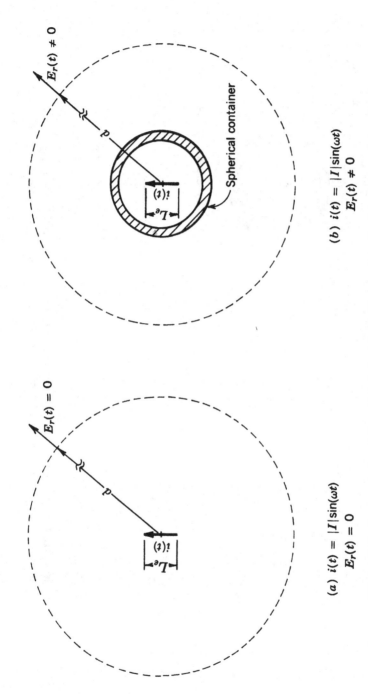

(a) $i(t) = |I|\sin(\omega t)$
$E_r(t) = 0$

(b) $i(t) = |I|\sin(\omega t)$
$E_r(t) \neq 0$

Figure 7-2. A current element's radiated electric field normal to the imaginary sphere of radius d is (a) zero when the element is in free space, and (b) nonzero when the element is in a closed, perfectly conducting, spherical container.

zero whenever it varies, that cannot happen instantaneously. The surface charge will be held to zero much of the time, but it will still cause radiation when it must be returned to zero. What is more, to return the outer surface charge to zero, there will have to be a current from the surface to the point at which it is grounded. Thus, there will be time-varying charges and currents on the outer surface of the container, and currents in the conductor connecting it to ground, all of which will radiate. This is illustrated in Fig. 7-3 for a current element centered in a perfectly conducting sphere the surface of which is connected to ground.

It should be clear, then, that full containment will generally not completely solve a radiation problem; it will alter the problem. Containment may bring about improvements in some cases, but in others it may worsen the problem. Thus, containment may sometimes be useful to maximize the reduction of unintentional radiations, but it is by no means a primary solution to the problem. Furthermore, in practice, the results expected from the nonspherical containment of the radiations of numerous current elements will be very difficult to predict.

Next considered is what to expect when containment can only be partially implemented.

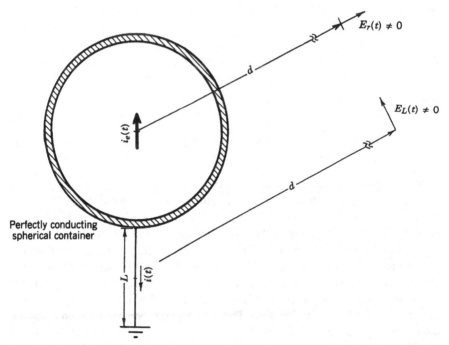

Figure 7-3. The radiated electric field in the direction normal to a grounded perfectly conducting sphere containing a current element would not be zero, nor would the field of the ground wire.

7.3 PARTIALLY CONTAINED RADIATING CURRENTS

Complete containment of radiating currents—placement in a highly conductive container with no openings—is generally difficult, if not impossible, to implement in practice. And openings in conductive containers will often cause stronger radiations than there would be with no container at all. To see that openings in highly conductive containers are capable of radiating as effectively as current-carrying conductors, or antennas, consider the following.

7.3.1   The Slot Element

The concept of an isolated short length of current, the current element, was introduced in Chapters 1 and 2, and from that concept descriptions of the radiations of numerous real antennas were obtained. Similarly, a *slot element*—a short, narrow opening containing a transverse electric field, in an infinite, perfectly-conducting plane—yields descriptions of the antenna-like behavior of longer slots in an infinite conducting plane.

The slot element and current element and their associated geometries are illustrated in Fig. 7-4 for comparison. The radiated electric field of the current $i(t) = |I| \sin(2\pi f t)$ from a current element of length L_e is given in Eq. 7-1. The radiated electric field of a slot element with the electric field $E(t) = |E| \sin(2\pi f t)$ across the slot's width w is

$$E_s(t) = E_s(\theta) \sin\left[\frac{2\pi f(t - d)}{c}\right]$$

where

$$E_s(\theta) = \frac{w|E|L_{s\lambda}}{2d} \sin\theta \qquad (7\text{-}2)$$

In these expressions, L_s is the slot length, $L_{s\lambda} = L_s f/c \leq \frac{1}{16}$, the relative sizes of L_s and w are such that $L_s > w \rightarrow 0$, and $d \gg L_s$ is the distance from the center of the slot to the observation point. The source of the current in a current element is not specified, nor is the source of the electric field in a slot element.

In comparing Eqs. 7-1 and 7-2, it is seen that radiation pattern factors of the slot element and the current element are identical except for the factors $w|E|L_{s\lambda}$ and $Z_0|I|L_{e\lambda}$. In Fig. 7-4 it is seen that the radiated electric field of a vertical slot element has the ϕ direction, whereas the radiated electric field of a vertical current element has the θ direction. Thus, at any common observation point the directions of their fields differ by 90°, but the relative magnitudes of their fields are the same. In other words, except in the conducting plane that contains it, the slot element has the same doughnut-shaped radiation pattern as the current element, as shown in Fig. 7-5. Thus, if $w|E|L_{s\lambda} = Z_0|I|L_{e\lambda}$, the radiations of

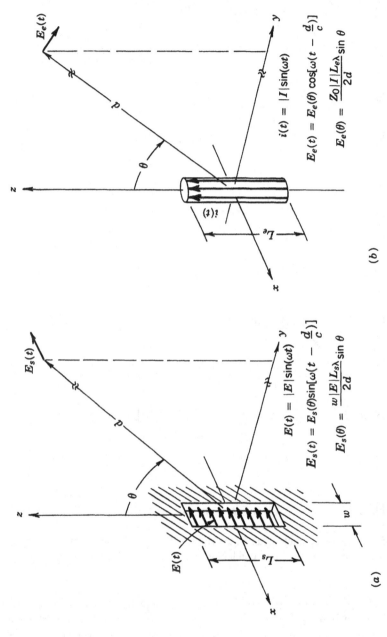

$E_s(t)$

z

p

θ

x

$E(t)$

L_s

w

y

$E(t) = |E|\sin(\omega t)$

$E_s(t) = E_s(\theta)\sin[\omega(t - \dfrac{d}{c})]$

$E_s(\theta) = \dfrac{w|E|L_s\lambda}{2d}\sin\theta$

(a)

$E_e(t)$

z

p

θ

x

$i(t)$

L_e

y

$i(t) = |I|\sin(\omega t)$

$E_e(t) = E_e(\theta)\cos[\omega(t - \dfrac{d}{c})]$

$E_e(\theta) = \dfrac{Z_0|I|L_e\lambda}{2d}\sin\theta$

(b)

Figure 7-4. The concept of (a) a slot element in an infinite conducting plane compared to that of (b) a current element in free space.

189

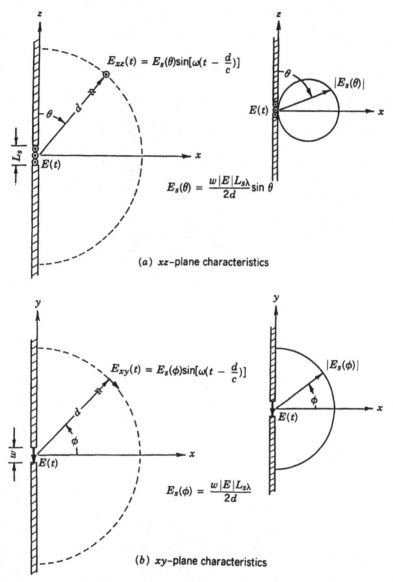

$$E_{xz}(t) = E_s(\theta)\sin[\omega(t - \tfrac{d}{c})]$$

$$E_s(\theta) = \frac{w|E|L_s\lambda}{2d}\sin\theta$$

(a) xz-plane characteristics

$$E_{xy}(t) = E_s(\phi)\sin[\omega(t - \tfrac{d}{c})]$$

$$E_s(\phi) = \frac{w|E|L_s\lambda}{2d}$$

(b) xy-plane characteristics

Figure 7-5. Radiation characteristics in the right half-space of a slot element along the z axis when the yz plane is a good conductor.

these two elementary sources will be the same except for the 90° difference in the directions of their fields.

The radiations of longer slots can be characterized with slot elements the same as the radiations of longer conductors were characterized with current elements. For example, the radiations of slots longer than a slot element can be described in the same way the radiations of current segments longer than current elements were described in Chapter 2. The results for a slot of length L on the z axis, obtained by replacing $Z_0 |I| L_{e\lambda}$ with $w|E| L_{s\lambda}$ in Eqs. 2-14 and 2-18, are

$$E_L(\theta) = \left[\frac{w|E|}{2\pi d} \sin \theta \right] \frac{\sin[\pi L_\lambda (\cos \theta - c/v_p)]}{\cos \theta - c/v_p} \tag{7.3a}$$

when the electric field propagates upward in the slot, and

$$E_L(\theta - \pi) = - \left[\frac{w|E|}{2\pi d} \sin \theta \right] \frac{\sin[\pi L_\lambda (\cos \theta + c/v_p)]}{\cos \theta + c/v_p} \tag{7-3b}$$

when the electric field propagates downward in the slot. Here, of course, v_p is the propagation velocity of the transverse electric field along the length of the slot.

The characteristics of radiating slots or unintentional slot antennas can now be examined.

7.3.2 Unintentional Slot Antennas

A slot in a conductor need not be intentionally excited for it to radiate. If an opening or a slot exists in any conductive enclosure that has surface currents on it, containment of the radiations causing those currents will seldom be achieved. Any container with a time-varying surface current on it, the direction of which is not parallel to a slot, will cause the slot to radiate much like an antenna. The fields induced in a slot by a current parallel to it are equal and opposite, as shown in Fig. 7-6, and cancel each other. However, a surface current in any other direction will have a component perpendicular to the slot, as shown in Fig. 7-7. And perpendicular currents will cause differing charge densities on opposite sides of a slot and electric fields that do not cancel.

Suppose the surface current density $J(t) = i(t)/m^2$ (amperes/meter^2) at some point along the edge of a slot is $J(t) = |J| \cos(2\pi f t)$, then the surface charge density at that point, in coulombs/meter^2, will be

$$\rho_s(t) = \int J(t) \, dt$$

$$= \int |J| \cos(2\pi f t) \, dt$$

$$= \frac{|J|}{2\pi f} \sin(2\pi f t) \tag{7-4a}$$

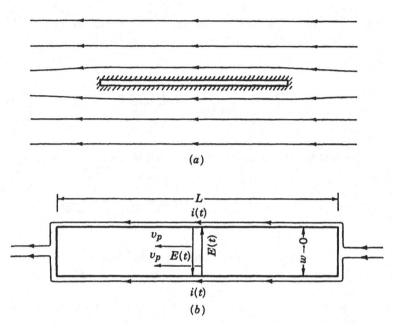

Figure 7-6. The pattern of (*a*) a surface current parallel to a slot and (*b*) the equal and opposite electric fields caused by equal charge densities on opposite sides of the slot.

That surface charge density causes an electric field normal to the slot edge that is

$$E(t) = \rho_s(t)/\epsilon$$

$$= \frac{|J|}{2\pi f \epsilon} \sin(2\pi f t)$$

$$= |E| \sin(2\pi f t) \qquad (7\text{-}4b)$$

where ϵ is the permittivity in the slot in farads/meter.

Referring to Fig. 7-8, it is seen that the electric field propagating to the right in the left half of the slot of length $2L$ causes the radiated electric field

$$E_{lr}(t) = E_L(\phi) \cos\left[2\pi f \left(t - \frac{L}{2v_p} - \frac{1}{c} \left(d + \frac{L}{2} \cos \phi \right) \right) \right]$$

$$= E_L(\phi) \cos\left[2\pi f \left(t - \frac{d}{c} \right) - \pi L_\lambda \left(\cos \phi + \frac{c}{v_p} \right) \right]$$

a distance $d + L/2 \cos \phi \cong d \gg 2L$ from the center of the slot. The electric

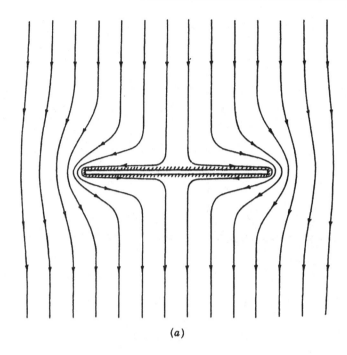

(a)

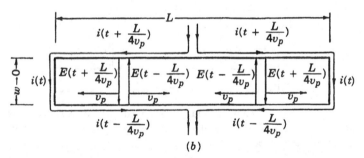

(b)

Figure 7-7. The pattern of (a) a surface current that is perpendicular to the slot on which it is incident and (b) the electric fields caused by the resulting charge densities on the sides of the slot.

field propagating to the right in the right half of the slot causes the radiated electric field

$$E_{rr}(t) = E_L(\phi) \cos \left[2\pi f \left(t + \frac{L}{2v_p} - \frac{1}{c} \left(d - \frac{L}{2} \cos \phi \right) \right) \right]$$

$$= E_L(\phi) \cos \left[2\pi f \left(t - \frac{d}{c} \right) + \pi L_\lambda \left(\cos \phi + \frac{c}{v_p} \right) \right]$$

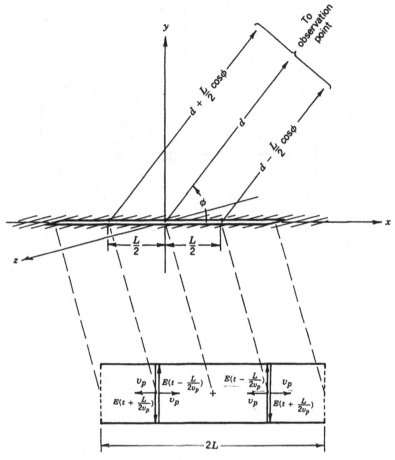

Figure 7-8. The radiation geometry and an expanded view of a slot and its electric fields when the slot is on the *x* axis and the conducting plane is the *xz* plane with a surface current in the *z* direction.

at the distance $d - L/2 \cos \phi \cong d$. Thus, recalling that $\cos(a - b) + \cos(a + b) = 2 \cos a \cos b$, the sum of those two fields at that observation point is seen to be

$$
E_{lr}(t) + E_{rr}(t) = E_L(\phi) \cos\left[2\pi f\left(t - \frac{d}{c} \right) - \pi L_\lambda \left(\cos \phi + \frac{c}{v_p} \right) \right]
$$

$$
+ E_L(\phi) \cos\left[2\pi f\left(t - \frac{d}{c} \right) + \pi L_\lambda \left(\cos \phi + \frac{c}{v_p} \right) \right]
$$

$$
= 2E_L(\phi) \cos\left[\pi L_\lambda \left(\cos \phi + \frac{c}{v_p} \right) \right] \cos\left[2\pi f\left(t - \frac{d}{c} \right) \right] \quad (7\text{-}5)
$$

Similarly, the electric field propagating to the left in the left half of the slot causes the radiated electric field

$$E_{ll}(t) = E_L(\phi - \pi) \cos\left[2\pi f\left(t + \frac{L}{2v_p} - \frac{1}{c}\left(d + \frac{L}{2}\cos\phi\right)\right)\right]$$

$$= E_L(\phi - \pi) \cos\left[2\pi f\left(t - \frac{d}{c}\right) - \pi L_\lambda\left(\cos\phi - \frac{c}{v_p}\right)\right]$$

at the same observation point. The electric field propagating to the left in the right half of the slot causes the radiated electric field

$$E_{rl}(t) = E_L(\phi - \pi) \cos\left[2\pi f\left(t - \frac{L}{2v_p} - \frac{1}{c}\left(d - \frac{L}{2}\cos\phi\right)\right)\right]$$

$$= E_L(\phi - \pi) \cos\left[2\pi f\left(t - \frac{d}{c}\right) + \pi L_\lambda\left(\cos\phi - \frac{c}{v_p}\right)\right]$$

at that observation point. Thus, the sum of the fields that propagate to the left in the slot is

$$E_{ll}(t) + E_{rl}(t) = E_L(\phi - \pi) \cos\left[2\pi f\left(t - \frac{d}{c}\right) - \pi L_\lambda\left(\cos\phi - \frac{c}{v_p}\right)\right]$$

$$+ E_L(\phi - \pi) \cos\left[2\pi f\left(t - \frac{d}{c}\right) + \pi L_\lambda\left(\cos\phi - \frac{c}{v_p}\right)\right]$$

$$= 2E_L(\phi - \pi) \cos[\pi L_\lambda(\cos\phi - c/v_p)] \cos[2\pi f(t - d/c)] \quad (7\text{-}6)$$

The total electric field radiated by the slot, therefore, is

$$E_S(t) = E_{lr}(t) + E_{rr}(t) + E_{ll}(t) + E_{rl}(t)$$

$$= E_S(\phi) \cos\left(2\pi f\left(t - \frac{d}{c}\right)\right)$$

where, using Eq. 7-3, with ϕ replacing θ because the slot is on the x axis,

$$E_S(\phi) = 2E_L(\phi) \cos \left[\pi L_\lambda \left(\cos \phi + \frac{c}{v_p} \right) \right]$$

$$+ 2E_L(\phi - \pi) \cos \left[\pi L_\lambda \left(\cos \phi - \frac{c}{v_p} \right) \right]$$

$$= \left[\frac{w|E|}{\pi d} \sin \phi \right] \frac{\sin[\pi L_\lambda(\cos \phi - (c/v_p))] \cos[\pi L_\lambda(\cos \phi + (c/v_p))]}{\cos \phi - (c/v_p)}$$

$$- \left[\frac{w|E|}{\pi d} \sin \phi \right] \frac{\sin[\pi L_\lambda(\cos \phi + (c/v_p))] \cos[\pi L_\lambda(\cos \phi - (c/v_p))]}{\cos \phi + (c/v_p)}$$

$$= \frac{w|E|}{\pi d} \sin \phi \frac{\cos \phi \sin[2\pi L_\lambda(c/v_p)] - (c/v_p)\sin(2\pi L_\lambda \cos \phi)}{[(c/v_p) - \cos \phi][(c/v_p) + \cos \phi)} \qquad (7\text{-}7)$$

The latter expression of Eq. 7-7 follows by applying the trigonometric identities $\sin(a \pm b) = \sin a \cos b \pm \sin b \cos a$ and $\cos(a \pm b) = \cos a \cos b \mp \sin a \sin b$.

The expression arrived at for the radiation pattern $E_S(\phi)$ in Eq. 7-7 is based on the assumption that the slot has length $2L \ll d$. The equivalent expression for the radiation pattern of any slot of length $L \ll d$ is

$$E_S(\phi) = \frac{w|E|}{\pi d} \sin \phi \frac{\cos \phi \sin(\pi L_\lambda(c/v_p)) - (c/v_p)\sin(\pi L_\lambda \cos \phi)}{((c/v_p) - \cos \phi)((c/v_p) + \cos \phi)} \qquad (7\text{-}8)$$

Assuming that $v_p \to c$, the radiation pattern for any slot of length $L \ll d$ is

$$E_S(\phi) = \frac{w|E|}{\pi d} \sin \phi \frac{\cos \phi \sin(\pi L_\lambda) - \sin(\pi L_\lambda \cos \phi)}{1 - \cos^2 \phi}$$

$$= \frac{w|E|}{\pi d} \frac{\cos \phi \sin(\pi L_\lambda) - \sin(\pi L_\lambda \cos \phi)}{\sin \phi} \qquad (7\text{-}9)$$

Radiation patterns obtained with Eq. 7-9 for slots of several different lengths L_λ in infinite conducting planes are shown in Fig. 7-9. The xy-plane radiation patterns are shown, but the pattern cross section is the same in any plane containing the x axis, except the xz plane, which is the conducting plane.

7.3.3 Partial Containment

A partial container of radiations is a highly conductive container with holes or openings in it that can in many cases be described as slots. Given the above descriptions of the radiations of slots in infinite planes it should be clear that any slot in a conducting surface is liable to be excited and radiate like an antenna,

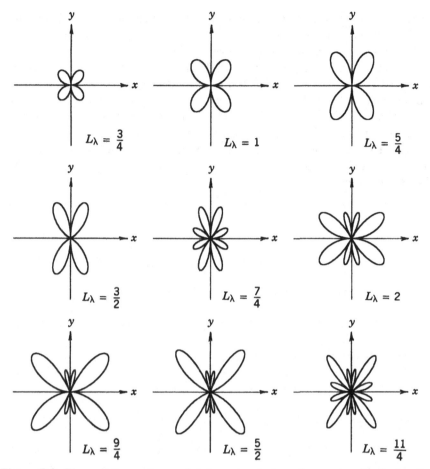

Figure 7-9. The radiation patterns of slots centered along the x axis with the lengths shown assuming the xz plane is infinite and perfectly conducting and has the same z-directed surface current in each case.

even though that surface is not infinite. Therefore, it should also be clear that partial containment of radiation may worsen the problem it is meant to solve.

7.4 SUMMARY AND CONCLUSIONS

Based on all of the above, it should be clear that containment is not the best way to reduce and control unintentional electromagnetic radiations. However, there are at least two more reasons that containment should be carefully considered before attempting to implement it.

The first of those reasons follows from the principle of reciprocity. That principle says that if electromagnetic radiations are reflected back toward their source they are very likely to interfere with the proper operation of that source. In other words, containment is highly likely to induce currents in the device contained that will corrupt its performance.

The second reason results from the difficulty in predicting the overall effects to expect from a particular container. Because of that difficulty, trial and error will generally be the only way to proceed. Each container will have to be implemented and its effectiveness measured. As a result, any attempt to contain radiations will generally be both time-consuming and costly.

The basic conclusion to be drawn from this chapter should be clear. Time and money will be better spent on efforts to control unwanted radiations rather than efforts to contain them.

CHAPTER 8

MEASUREMENT OF UNINTENTIONAL RADIATIONS

8.1 INTRODUCTION

To determine the extent to which unintentional electromagnetic radiations must be reduced, they must first be measured. The primary purpose of this chapter is to make known what must be understood, and what must be done, for those measurements to be performed accurately and effectively. The measurement environment, the measurement setup, and certain aspects of the measurement procedure, all prescribed by various national and international agencies are described and analyzed. The basic theory behind the methods specified and those aspects of the measurements on which their accuracy largely depend are the topics of primary concern.

To measure electromagnetic radiations accurately, the receiving antennas used must be accurately calibrated, and the test site used must accurately simulate the ideal measurement environment. Therefore, attention will be focused here on (1) the characteristics of receiving antennas and how they can be accurately calibrated, (2) the characteristics of the ideal measurement environment and how its simulation by a given test site can be accurately validated, and (3) the appropriate use of both antennas and test sites in performing all of the measurements required, both to validate the test site and to minimize unintentional radiations.

Because they are used both for test site validation, and to search for and measure unintentional electromagnetic radiations, receiving antennas are first to be discussed.

8.2 CHARACTERISTICS OF RECEIVING ANTENNAS

The primary objective of this section is to derive accurate descriptions of those characteristics of receiving antennas that are significant in accurately performing the required measurements. As usual, however, antenna characteristics that are important in practice have originated in theory. Therefore, the conceptual tool from which much of antenna theory arises—the current element—will be the first "antenna" to be discussed.

8.2.1 Effective Length

The concept of the *effective length* of an antenna derives from the definition of the current element. As shown in Fig. 8-1, a current element of length L_e that is centered along the z-axis and has the current $i(t) = |I| \sin(\omega t)$, would have the radiated electric field

$$E_e(t) = E_e(\theta) \cos\left[\omega\left(t - \frac{d}{c}\right)\right]$$

where

$$E_e(\theta) = \frac{Z_0|I|}{2\lambda d} L_e \sin\theta$$

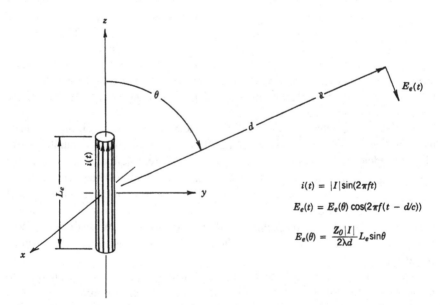

$$i(t) = |I|\sin(2\pi f t)$$

$$E_e(t) = E_e(\theta)\cos(2\pi f(t - d/c))$$

$$E_e(\theta) = \frac{Z_0|I|}{2\lambda d} L_e \sin\theta$$

Figure 8-1. The radiated electric field of a current element centered on the z axis.

from which it follows that

$$L_e \sin \theta = \frac{2\lambda d}{Z_0} \frac{E_e(\theta)}{|I|} \tag{8-1}$$

From this relationship the effective length of a current element when it is *transmitting* is defined to be

$$L_{ee}^T(\theta) \equiv \frac{2\lambda d}{Z_0} \frac{E_e(\theta)}{|I|}$$
$$= L_e \sin \theta \tag{8-2}$$

Suppose, as in Fig. 8-2, the same current element were to receive an electric field $E(t)$ from the θ direction. The voltage $V_e(\theta)$ would be induced over its length to establish a field equal and opposite to $E(t) \sin \theta$, thus satisfying the boundary conditions at the surface of the current element. Therefore,

$$V_e(\theta) = [E(t) \sin \theta] L_e \tag{8-3}$$

and the effective length of a current element when it is *receiving* is defined to be

$$L_{ee}^R(\theta) \equiv \frac{V_e(\theta)}{E(t)} = L_e \sin \theta \tag{8-4}$$

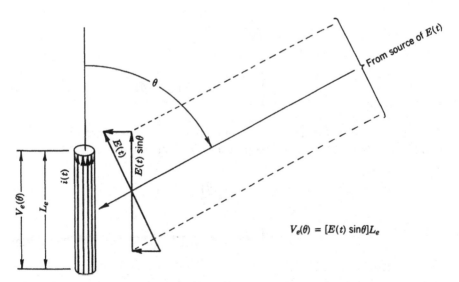

Figure 8-2. The voltage $V_e(\theta)$ is induced over the length of a current element by the incident electric field $E(t)$ when the angle of incidence is θ.

The effective length of any antenna can be similarly defined. The radiated electric field of an antenna with the radiation pattern factor $E_a(\theta)$ has the general description

$$E_a(t) = E_a(\theta) \cos\left[\omega\left(t - \frac{d}{c}\right)\right]$$ (8-5)

Therefore, similar to the current element, the effective length of any antenna when it is *transmitting* the electric field $E_a(t)$ is defined to be

$$L_{eff}^T(\theta) \equiv \frac{2\lambda d}{Z_0} \frac{E_a(\theta)}{|I_a|}$$ (8-6)

where $|I_a|$ is the amplitude of the sinusoidal antenna current. The effective length of any antenna when it is *receiving* an electric field $E(t)$ is defined to be

$$L_{eff}^R(\theta) \equiv \frac{V_a(\theta)}{E(t)}$$ (8-7)

where $V_a(\theta)$ is the amplitude of the antenna response voltage.

Now, suppose the field of a current element with a current $|I_e| \sin(\omega t)$ were to excite an antenna with an effective length $L_{eff}^R(\theta)$ in the plane of the current element. Then, from the current element to the antenna the response-to-excitation ratio would be

$$\frac{V_a(\theta)}{|I_e|} = \frac{E_e(t)L_{eff}^R(\theta)}{|I_e|}$$

$$= \left[\frac{Z_0}{2\lambda d} L_e \sin\theta\right] L_{eff}^R(\theta)$$ (8-8)

And, if the field of the antenna with a current $|I_a| \sin(\omega t)$ were to excite the current element, then the response-to-excitation ratio would be

$$\frac{V_e(\theta)}{|I_a|} = \frac{E_a(t)L_{ee}^R(\theta)}{|I_a|}$$

$$= \left[\frac{Z_0}{2\lambda d} L_{eff}^T(\theta)\right] L_e \sin\theta$$ (8-9)

From the principle of reciprocity it is known that these response-to-excitation ratios would be equal. Therefore,

$$\frac{V_a(\theta)}{|I_e|} = \left[\frac{Z_0}{2\lambda d} L_e \sin \theta \right] L_{eff}^R(\theta)$$

$$= \left[\frac{Z_0}{2\lambda d} L_{eff}^T(\theta) \right] L_e \sin \theta = \frac{V_e(\theta)}{|I_a|} \tag{8-10}$$

from which it is seen that

$$L_{eff}^T(\theta) = L_{eff}^R(\theta) \tag{8-11}$$

Thus, like a current element, the effective length of any antenna is the same whether it is transmitting or receiving.

8.2.2 Dipole Effective Lengths

It was seen in Chapter 2 that a center-fed, transmitting dipole antenna of length L has the radiated electric field

$$E_D(t) = E_D(\theta) \sin \left(\omega \left(t - \frac{d}{c} \right) \right)$$

where

$$E_D(\theta) = \frac{Z_0 |I|}{\pi d} \frac{\cos(\pi L_\lambda) - \cos(\pi L_\lambda \cos \theta)}{\sin \theta} \tag{8-12}$$

In this expression, $|I|$ is the amplitude of both the applied current and the reflected current. Those currents together make up the total current of amplitude $|I_0|$ that exists at the center of the dipole.

Therefore, assuming a dipole is centered on the z-axis, at a distance z from its center where $-L/2 \leq z \leq L/2$, the total current is the difference between the applied current

$$i_a(t, z) = |I| \sin \left(\omega \left(t + \frac{L - 2z}{2v_p} \right) \right)$$

$$= |I| \sin(\omega t + \pi L_\lambda - 2\pi z_\lambda) \tag{8-13a}$$

and the reflected current

$$i_r(t, z) = |I| \sin\left(\omega\left(t - \frac{L - 2z}{2v_p}\right)\right)$$

$$= |I| \sin(\omega t - \pi L_\lambda + 2\pi z_\lambda) \qquad (8\text{-}13b)$$

If follows from these equations, then, that the total current at the center of the dipole, where $z = 0$, will be

$$i_a(t, 0) - i_r(t, 0) = |I| \sin(\omega t + \pi L_\lambda) - |I| \sin(\omega t - \pi L_\lambda)$$

$$= 2|I| \sin(\pi L_\lambda) \cos(\omega t)$$

$$= |I_0| \cos(\omega t) \qquad (8\text{-}14)$$

This follows from the identity $2 \sin y \cos x = \sin(x + y) - \sin(x - y)$, by letting $x = \omega t$ and $y = \pi L_\lambda$.

It follows from Eqs. 8-12, 8-13, and 8-14, that if the amplitude of the current at the center of a dipole of length L is $|I_0| = 2|I| \sin(\pi L_\lambda)$, its radiation pattern factor will be

$$E_D(\theta) = \frac{Z_0|I|}{\pi d} \frac{\cos(\pi L_\lambda) - \cos(\pi L_\lambda \cos \theta)}{\sin \theta}$$

$$= \frac{Z_0|I_0|}{2d} \frac{\cos(\pi L_\lambda) - \cos(\pi L_\lambda \cos \theta)}{\pi \sin(\pi L_\lambda) \sin \theta}$$

$$= \frac{Z_0|I_0|}{2\lambda d} L \left[\frac{\cos(\pi L_\lambda) - \cos(\pi L_\lambda \cos \theta)}{\pi L_\lambda \sin(\pi L_\lambda) \sin \theta} \right]$$

$$= \frac{Z_0|I_0|}{2\lambda d} L_{eff}(\theta) \qquad (8\text{-}15)$$

In other words, the effective length of any *center-fed* dipole of length L is

$$L_{eff}(\theta) = L \left[\frac{\cos(\pi L_\lambda) - \cos(\pi L_\lambda \cos \theta)}{\pi L_\lambda \sin(\pi L_\lambda) \sin \theta} \right] \qquad (8\text{-}16)$$

The effective lengths of short dipoles and half-wave dipoles, are now obtained using the above results.

8.2.2.1 Short Dipoles
The effective length of short dipoles, the lengths of which are $L_\lambda \leq \frac{1}{10}$, can be found as follows. Since the length of any short dipole will be $L_\lambda \leq \frac{1}{10}$, it follows that $\pi L_\lambda/2 \leq \pi/20$, from which it further follows that $\sin(\pi L_\lambda/2) \cong \pi L_\lambda/2$ is a well-justified approximation. From several previously

used trigonometric identities given in Appendix A, it can also be seen that when $x \le 20$, $1 - \cos(2x) = 1 - \cos^2(x) + \sin^2(x) = 2\sin^2(x)$, and $2\sin^2(x) \cong 2x^2$. Based on these observations and approximations, it follows from Eq. 8-16 that the effective length of a short dipole will be

$$
\begin{aligned}
L_{eff}(\theta) &= -L\left[\frac{\cos(\pi L_\lambda \cos\theta) - \cos(\pi L_\lambda)}{\pi L_\lambda \sin(\pi L_\lambda)\sin\theta}\right] \\
&\cong -L\left\{\frac{1 - \cos(\pi L_\lambda) - 2\sin^2[(\pi L_\lambda/2)\cos\theta]}{(\pi L_\lambda)^2 \sin\theta}\right\} \\
&\cong -L\left[\frac{2(\pi L_\lambda/2)^2 - 2(\pi L_\lambda/2)^2\cos^2\theta}{(\pi L_\lambda)^2 \sin\theta}\right] \\
&= -\frac{L}{2}\left(\frac{1 - \cos^2\theta}{\sin\theta}\right) \\
&= -\frac{L}{2}\sin\theta
\end{aligned}
\tag{8-17}
$$

From this it follows that the radiated power density coming from a short dipole observed at a distance $d \gg \lambda$ will be

$$
\begin{aligned}
P_t(\theta) &= \frac{|E_D(\theta)|^2}{Z_0} = \frac{Z_0|I_0|^2}{4\lambda^2 d^2}L_{eff}^2(\theta) \\
&= \frac{Z_0|I_0|^2 L^2}{16\lambda^2 d^2}\sin^2\theta
\end{aligned}
\tag{8-18}
$$

And the total radiated power W_t over a sphere of radius $r = d \gg \lambda$, which has the dipole at its center, will be the integral of $P_t(\theta)$ over the surface of the sphere. Thus, the total power radiated by a short dipole will be

$$
\begin{aligned}
W_t &= \int_0^{2\pi}\left[\int_0^\pi \frac{|E_D(\theta)|^2}{Z_0}r\sin\theta\,d\theta\right]r\,d\phi \\
&= \frac{2\pi d^2}{Z_0}\int_0^\pi |E_D(\theta)|^2 \sin\theta\,d\theta \\
&= \frac{2\pi d^2}{Z_0}\left(\frac{Z_0^2|I_0|^2 L^2}{16\lambda^2 d^2}\right)\int_0^\pi \sin^3\theta\,d\theta \\
&= \frac{\pi Z_0|I_0|^2 L^2}{6\lambda^2}
\end{aligned}
\tag{8-19}
$$

Assuming there are no losses in the dipole, the total radiated power will also have the value

$$W_t = |I_0|^2 R_D \tag{8-20}$$

where R_D is the radiation resistance of the dipole seen from the terminals at its center. From Eqs. 8-19 and 8-20, then, it follows that the radiation resistance of a short dipole, for which $L_\lambda \le \frac{1}{10}$, is

$$R_D = \frac{\pi Z_0 L^2}{6\lambda^2} = 20\pi^2 L_\lambda^2. \tag{8-21}$$

8.2.2.2 The Half-Wave Dipole   A half-wave dipole of length L, with the current $i_0(t) = |I_0| \cos(\omega t)$ at its center, has the radiated electric field

$$E_D(t) = E_D(\theta) \sin\left[\omega \left(t - \frac{d}{c} \right) \right]$$

and, since $L_\lambda = \frac{1}{2}$, from Eq. 8-16 its effective length is

$$
\begin{aligned}
L_{eff}(\theta) &= L \left[\frac{\cos(\pi/2) - \cos((\pi/2)\cos\theta)}{(\pi/2)\sin(\pi/2)\sin\theta} \right] \\
&= -\frac{2L}{\pi} \frac{\cos((\pi/2)\cos\theta)}{\sin\theta} \\
&= -\frac{\lambda}{\pi} \frac{\cos((\pi/2)\cos\theta)}{\sin\theta}
\end{aligned}
\tag{8-22}
$$

Therefore,

$$
\begin{aligned}
E_D(\theta) &= \frac{Z_0 |I_0|}{2\lambda d} L_{eff}(\theta) \\
&= -\frac{Z_0 |I_0|}{2\pi d} \left[\frac{\cos((\pi/2)\cos\theta)}{\sin\theta} \right]
\end{aligned}
\tag{8-23}
$$

Thus $P_t(\theta)$, the radiated power density a distance $d \gg \lambda$ from a radiating half-wave dipole, will be

$$P_t(\theta) = \frac{E_D^2(\theta)}{Z_0} = \frac{Z_0|I_0|^2}{4\lambda^2 d^2} L_{eff}^2(\theta)$$

$$= \frac{Z_0|I_0|^2 L^2}{\pi^2 \lambda^2 d^2} \left\{ \frac{\cos[(\pi/2)\cos\theta]}{\sin\theta} \right\}^2$$

$$= \frac{Z_0|I_0|^2}{4\pi^2 d^2} \left\{ \frac{\cos[(\pi/2)\cos\theta]}{\sin\theta} \right\}^2 \tag{8-24}$$

The total power radiated is found, as before, by evaluating the integral of $P_t(\theta)$ over a sphere of radius $d \gg \lambda$. However, in this case, a close approximation of $P_t(\theta)$ will be used to simplify the mathematics required to evaluate the integral of $\cos^2[(\pi/2)\cos\theta]/\sin\theta$. Note from Fig. 8-3 that, for all practical purposes,

$$0.43\sin^5\theta + 0.57\sin^3\theta = \frac{\cos^2[(\pi/2)\cos\theta]}{\sin\theta} \tag{8-25}$$

Since for $n > 1$,

$$\int_0^\pi \sin^n\theta \, d\theta = \frac{n-1}{n} \int_0^\pi \sin^{n-2}\theta \, d\theta$$

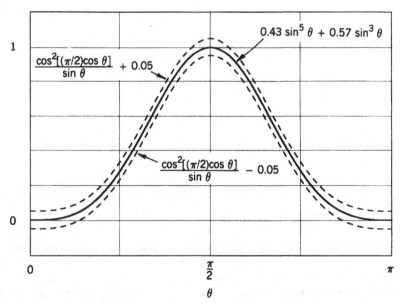

Figure 8-3. A graphical comparison of the function $\cos^2[(\pi/2)\cos\theta]/\sin\theta$, and its approximation $0.43\sin^5\theta + 0.57\sin^3\theta$.

it follows that

$$W_t = \int_0^{2\pi} \left[\int_0^\pi \frac{|E_D(\theta)|^2}{Z_0} \, r \sin\theta \, d\theta \right] r \, d\phi$$

$$= \frac{2\pi d^2}{Z_0} \int_0^\pi |E_D(\theta)|^2 \sin\theta \, d\theta$$

$$= 2\pi d^2 \left(\frac{Z_0 |I_0|^2}{4\pi^2 d^2} \right) \int_0^\pi \frac{\cos^2[(\pi/2)\cos\theta]}{\sin\theta} \, d\theta$$

$$\cong \left(\frac{Z_0 |I_0|^2}{2\pi} \right) \left(0.43 \int_0^\pi \sin^5\theta \, d\theta + 0.57 \int_0^\pi \sin^3\theta \, d\theta \right)$$

$$= \frac{Z_0 |I_0|^2}{2\pi} \left(0.43 \frac{16}{15} + 0.57 \frac{4}{3} \right)$$

$$= \frac{0.61 Z_0}{\pi} |I_0|^2 \tag{8-26}$$

Based on this result, and recalling that the total power radiated equals the product

$$W_t = R_D |I_0|^2 \tag{8-27}$$

it follows that the radiation resistance of the half-wave dipole is

$$R_D = \frac{0.61 Z_0}{\pi} = 73.2 \text{ ohms} \tag{8-28}$$

8.3   THE ANTENNA FACTOR

The basic measurement system required to measure electromagnetic radiations consists of a receiving antenna, a balun (*bal*anced to *un*balanced transformer) and matching network, a coaxial cable, and a measuring receiver, or spectrum analyzer. These components of the basic measurement system are connected to one another and generally positioned above and below the ground plane of the test site, as illustrated in Fig. 8-4. Their function is to develop and present to the observer a response at the measuring receiver that is directly related to the electric field strength arriving at the receiving antenna. That relationship is described by the *antenna factor*, which is determined by the antenna's effective length, and by any power losses that occur between the antenna and receiver.

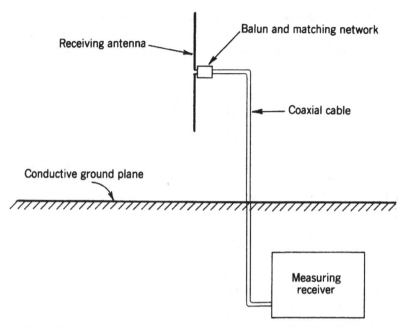

Figure 8-4. The basic measurement system and typical positioning of its components on the test site.

More specifically, suppose that $|E_R|$ is the magnitude of the electric field arriving at the antenna. $|E_R|$ is in the same plane as the antenna and it is normal to and arrives from the θ direction. If $|V_R(\theta)|$ is the magnitude of the voltage response of the measuring receiver, then the antenna factor of the measurement system is

$$AF(\theta) = \frac{|E_R|}{|V_R(\theta)|} \tag{8-29}$$

Thus, if the antenna factor $AF(\theta)$ is known and the voltage response $|V_R(\theta)|$ is measured and displayed, then the measured value of electric field strength arriving at the antenna is

$$|E_R| = AF(\theta)|V_R(\theta)| \tag{8-30}$$

It is clear from this equation that an accurate knowledge of the antenna factor $AF(\theta)$ is essential to the accurate determination of the electric field strength $|E_R|$. Therefore, the basic parameters that enter into the antenna factor are now identified.

A simple equivalent network of the basic measurement system is illustrated in Fig. 8-5. In that network, $Z_a = R_a + jX_a$ is the antenna impedance, as seen from the terminals that connect the antenna to the balun and matching network. And $V_a(\theta)$ is the *open-circuit* output voltage of the receiving antenna that is available at those same terminals. Thus, the power available from the antenna is

$$W_a(\theta) = \frac{|V_a(\theta)|^2}{4R_a} \tag{8-31}$$

The antenna side of the balun and matching network has an input impedance $Z_b = R_b + jX_b$. Therefore, $W_b(\theta)$, the power delivered to the balun and matching network by the antenna, will be

$$W_b(\theta) = \frac{|V_a(\theta)|^2 R_b}{|Z_a + Z_b|^2} \tag{8-32}$$

Thus, the ratio of the power available from the antenna to the power delivered to the balun and matching network will be

$$\frac{W_a(\theta)}{W_b(\theta)} = \frac{|Z_a + Z_b|^2}{4R_a R_b} \tag{8-33}$$

It is clear that $W_a(\theta) \geq W_b(\theta)$, and that $W_b(\theta) = W_a(\theta)$ only if the impedances Z_a and Z_b are such that $Z_b = Z_a^* = R_a - jX_a$. In other words, if the impedances Z_a and Z_b are complex conjugates, then all of the available power from the antenna is delivered to the balun and matching network. Otherwise, there is said to be a mismatch loss between the two. The power loss due to any mismatch between Z_a and Z_b is therefore

$$M_{ab}^2 = \frac{|Z_a + Z_b|^2}{4R_a R_b} \geq 1 \tag{8-34}$$

Mismatch losses can also occur at the junctions between the balun and matching network and the cable, and between the cable and the receiver. However, it is assumed here that any mismatches at those junctions have been eliminated.

As shown in Fig. 8-5, K_b^2 is the power loss through the balun and matching network, and $\epsilon^{2\alpha\ell}$ is the power loss through the coaxial cable, where α is the attenuation factor of the cable, and ℓ is its length. Therefore, if $W_R(\theta)$ is the power delivered to the measuring receiver, and the only impedance mismatch in the system occurs between the antenna and the balun, then

$$W_a(\theta) = M_{ab}^2 K_b^2 \epsilon^{2\alpha\ell} W_R(\theta) \tag{8-35}$$

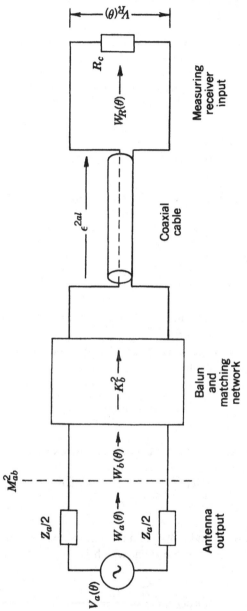

Figure 8-5. An equivalent network of the basic measurement system showing the power losses from the receiving antenna to the measuring receiver.

Assuming there is no mismatch between the cable and the receiver input, the receiver input impedance equals R_c, the characteristic impedance of the cable. From this it follows that the power delivered to the measuring receiver is

$$W_R(\theta) = \frac{|V_R(\theta)|^2}{R_c} \tag{8-36}$$

Given the expression for $W_a(\theta)$ of Eq. 8-31, it is seen from Eqs. 8-35 and 8-36 that

$$\frac{W_a(\theta)}{W_R(\theta)} = M_{ab}^2 K_b^2 \epsilon^{2\alpha\ell} = \frac{|V_a(\theta)|^2 R_c}{|V_R(\theta)|^2 4R_a} \tag{8-37}$$

Finally, taking the square root of both sides of the latter of these equations yields

$$\frac{|V_a(\theta)|}{|V_R(\theta)|} = 2\sqrt{\frac{R_a}{R_c}} \, M_{ab} K_b \epsilon^{\alpha\ell} = \frac{|Z_a + Z_b| K_b \epsilon^{\alpha\ell}}{\sqrt{R_b R_c}} \tag{8-38}$$

This is the ratio of the open-circuit output voltage of the antenna to the measured voltage response of the receiver.

From previous discussions, it is known that the magnitude of the effective length of any receiving antenna is

$$|L_{eff}(\theta)| \equiv \frac{|V_a(\theta)|}{|E_R|} \tag{8-39}$$

Therefore, from the previous two equations, it follows that

$$\frac{|E_R|}{|V_R(\theta)|} = \frac{1}{|L_{eff}(\theta)|} \frac{|V_a(\theta)|}{|V_R(\theta)|} \tag{8-40}$$

Together, Eqs. 8-29, 8-38, and 8-40, yield the general expression for the antenna factor of the basic measurement system, which is

$$AF(\theta) = \frac{|Z_a + Z_b| K_b \epsilon^{\alpha\ell}}{\sqrt{R_b R_c} \, |L_{eff}(\theta)|} \tag{8-41}$$

It is clear that any antenna factor depends on θ, the angle of incidence of a received wave, which will vary. It is also clear that the antenna factor depends upon the antenna impedance Z_a, which in free space is a constant. However, in the measurement environment specified here, Z_a will not be constant, and, as a result, $AF(\theta) = AF(\theta, Z_a)$. The impedance variations to expect from a half-wave dipole antenna are given and discussed below.

8.4 THE MEASUREMENT ENVIRONMENT

When unintentional electromagnetic emissions were first becoming recognized as a problem, their measurements were generally made in open fields, in empty parking lots, or on flat rooftops. There was nothing very sophisticated about test sites used in those days, so long as they were reasonably flat and relatively free of reflecting objects. Today, however, it is generally agreed that measurements should be made on test sites that are valid simulations of an infinite, empty, half-space, bounded by a perfectly conducting plane of infinite extent.

While this is a more precise definition, it is, of course, a mathematical ideal that can only be approximated in practice. Therefore, the acceptability of any given test site is determined by a comparison of measured responses obtained on the test site to calculated responses obtained for the ideal test site. If the results of those comparisons are sufficiently close to one another, then the test site in question is considered to provide an acceptable measurement environment.

8.4.1 Test-Site Construction

To closely approximate the behavior of the ideal test site, the surface over which the measurements are to be made must be highly conductive. Thus, the earth, or other surface, over which the measurements are to be made is usually covered with sheet metal or a metal screen of small mesh. Because the infinite conductive plane of the ideal environment would be a perfect electrical ground, the sheet metal or metal screen must also be well grounded. As a result, in theory, and in practice, the conductive plane over which the measurements are to be made is generally referred to as the *ground plane*.

It is also generally agreed that the ground plane of an actual test site must extend under and beyond both the equipment being tested and the measurement antenna. And it is agreed that the ground plane must cover the area between the equipment being tested and the measurement antenna. The open "area" required of a test site is the space above and around the ground plane. In the region above the ground plane there must be no objects other than the receiving antenna and its cable and the equipment under test. There can be nothing near the ground plane that might reflect or otherwise interfere with the electromagnetic radiations that are to be measured. The objective, of course, is to simulate an empty half-space, the emptiness of which is infinite in extent. However, for actual test sites, neither the overall extent of the metal ground plane nor the extent of the open area is generally agreed on. Because of that, rather than being based on specifics of its construction, a test site's acceptability is based almost entirely on its behavior.

How to scrutinize a test site's behavior to determine whether or not it satis-factorily simulates the ideal test site is the next subject of discussion.

8.4.2 Test-Site Validation

As mentioned above, the acceptability of open area test sites is primarily deter-mined by comparing *measured* responses obtained on those test sites with *calcu-lated* responses obtained assuming the same conditions on the ideal test site. To make these measurements and calculations accurately, the performance charac-teristics of the ideal test site and those of the antennas used must both be well known. Therefore, it is generally agreed that the half-wave dipole should be the standard antenna used both in making the measurements and in making the calculations. The performance characteristics of the half-wave dipole are well known and widely documented.

When a half-wave dipole is the receiving antenna, then the antenna impedance will be $Z_a = Z_D = 73.2 + j42.5$ ohms and its effective length will be $|L_{eff}(\theta)| = (\lambda/\pi)\{\cos[(\pi/2)\cos\theta]/\sin\theta\}$, for all θ from 0 to π. And, because it will be shown that Z_D will vary, the antenna factor of the measurement system with a half-wave dipole as the receiving antenna will be given as

$$AF(\theta, Z_D) = \frac{\pi K_b \epsilon^{\alpha\ell}}{\lambda\sqrt{R_b R_c}} \frac{|Z_D + Z_b|\sin\theta}{\cos[(\pi/2)\cos\theta]} \qquad (8\text{-}42)$$

This expression for the antenna factor of the basic measurement system, with a half-wave dipole for the receiving antenna, will be the antenna factor used in all of the examples that follow.

8.4.2.1 The Quantity Measured The primary measurement that is made to ascertain whether or not a given test site behaves acceptably is the measurement of *site attenuation*, which is an insertion loss. The two basic steps required to measure site attenuation are illustrated in Fig. 8-6. The first step is to measure the power loss through the measurement system with the baluns, or in some cases the cables, of both antennas directly connected together. The second step is to "insert" the antennas and the test site and again measure the power loss using specified measurement geometries. The output of the signal generator is kept the same for both measurements, and the receiving antenna height is varied to find the minimum loss in the second step of the measurements.

In both steps, the output voltage of the signal generator $|V_G|$ is the input to the transmitting antenna cable. And $|V_R|$ and $|V'_R|$ are the measuring receiver voltage responses before, and after insertion. Thus, the site attenuation insertion loss at any frequency will be the ratio of W_R, the power delivered to the mea-suring receiver before insertion, to W'_R, the power delivered to the measuring receiver after insertion. Alternative expressions for that power loss ratio are

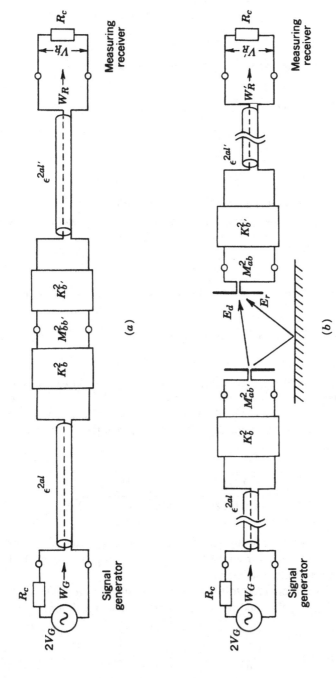

Figure 8-6. Equivalent networks of the basic measurement system for the site attenuation measurement: (a) before and (b) after test-site insertion.

$$\frac{W_R}{W'_R} = \frac{|V_R|^2/R_c}{|V'_R|^2/R_c} = \frac{|V_R|^2}{|V'_R|^2} \tag{8-43}$$

The value of the voltage $|V_R|$ can be found as follows. If $|V_G|^2/R_c$ is the power delivered to the cable of the transmitting antenna, then

$$\frac{|V_G|^2}{R_c} = \epsilon^{2\alpha\ell} K_b^2 M_{bb'}^2 K_{b'}^2 \epsilon^{2\alpha\ell'} \frac{|V_R|^2}{R_c} \tag{8-44}$$

where

$\epsilon^{2\alpha\ell}$ and $\epsilon^{2\alpha\ell'}$ = the power losses through the antenna cables
K_b^2 and $K_{b'}^2$ = the power losses through the balun and matching networks
$M_{bb'}^2$ = the mismatch loss between the baluns
$|V_R|^2/R_c$ = the power delivered to the measuring receiver

Thus, the voltage seen at the measuring receiver *before* insertion is

$$|V_R| = \frac{|V_G|}{\epsilon^{\alpha\ell} K_b M_{bb'} K_{b'} \epsilon^{\alpha\ell'}} \tag{8-45}$$

The voltage seen at the measuring receiver *after* insertion is found in terms of the antenna factors of the measurement system. If $|E_R(\theta)|$ is the electric field strength arriving at the receiving antenna, and $AF_R(\theta, Z_a)$ is the antenna factor of the receiving antenna side of the system, then the voltage seen by the measuring receiver is

$$|V'_R| = \frac{|E_R(\theta)|}{AF_R(\theta, Z_a)} = \frac{Z_0|I_0|}{2\lambda d} \frac{|L_{eff}^T(\theta)|}{AF_R(\theta, Z_a)} \tag{8-46}$$

The current $|I_0|$ in this expression is the current at the center of the transmitting antenna, which results from the signal generator output voltage $|V_G|$ being applied to the transmitting antenna cable.

Thus, the power delivered to the transmitting antenna from the signal generator is $|I_0|^2 R_a$, and the total power delivered by the signal generator is

$$\frac{|V_G|^2}{R_c} = \epsilon^{2\alpha\ell} K_b^2 M_{ab}^2 |I_0|^2 R_a \tag{8-47}$$

Therefore,

$$|I_0| = \frac{|V_G|}{\epsilon^{\alpha \ell} K_b M_{ab} \sqrt{R_c R_a}} = |V_G| \frac{2\sqrt{R_b/R_c}}{\epsilon^{\alpha \ell} K_b |Z_a + Z_b|} \qquad (8\text{-}48)$$

and Eqs. 8-46, 8-48, and 8-41, in that order, imply that

$$\frac{|V_R'|}{|V_G|} = \frac{Z_0}{2\lambda d} \left[\frac{2\sqrt{R_b/R_c}}{\epsilon^{\alpha \ell} K_b |Z_a + Z_b|} \right] \frac{|L_{eff}^T(\theta)|}{AF_R(\theta)}$$

$$= \frac{Z_0}{\lambda R_c d} \left[\frac{\sqrt{R_b R_c} |L_{eff}^T(\theta)|}{\epsilon^{\alpha \ell} K_b |Z_a + Z_b|} \right] \frac{1}{AF_R(\theta)}$$

$$= \frac{Z_0}{\lambda R_c d} \frac{1}{AF_T(\theta)} \frac{1}{AF_R(\theta)} \qquad (8\text{-}49)$$

Thus, it is seen from Eqs. 8-45 and 8-49 that site attenuation can be expressed as the power ratio

$$\frac{W_R}{W_R'} = \frac{|V_R|^2}{|V_R'|^2} = \frac{\lambda^2 R_c^2 d^2}{Z_0^2} \frac{AF_R^2(\theta, Z_D) AF_T^2(\theta, Z_D)}{\epsilon^{2\alpha \ell} K_b^2 M_{bb'}^2 K_{b'}^2 \epsilon^{2\alpha \ell'}} \qquad (8\text{-}50)$$

and it can also be expressed as the voltage ratio

$$\frac{|V_R|}{|V_R'|} = \frac{\lambda R_c d}{Z_0} \frac{AF_R(\theta, Z_D) AF_T(\theta, Z_D)}{\epsilon^{\alpha \ell} K_b M_{bb'} K_{b'} \epsilon^{\alpha \ell'}} \qquad (8\text{-}51)$$

The above equations describe the site attenuation that would be measured in free space when the antennas are separated by the distance d. On actual test sites, however, there are two additional factors that must be considered. They are the presence of the ground plane and the effects of measurement geometry.

8.4.2.2 Environment, Geometry, and Measured Radiations As previously discussed in Chapter 6 (see Fig. 6-28 and the associated discussion), the effect of a ground plane is to establish radiating images of any radiating currents that are positioned over it. Therefore, the effect of a ground plane in the environment prescribed for measuring electromagnetic radiations is to establish radiating images of any antennas that are placed in that environment. Vertical and horizontal dipoles positioned over a test-site ground plane and their images are illustrated in Fig. 8-7. The overall measurement geometry for the site attenuation measurement, including both the source and receiving antennas and the effect of the image of the source antenna, is illustrated in Fig. 8-8.

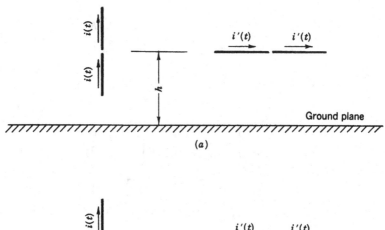

(a)

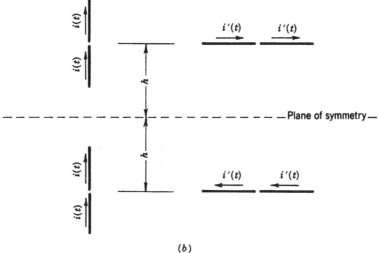

(b)

Figure 8-7. The radiated electromagnetic fields over a highly conductive ground plane are the sums of the radiations of (a) the antennas over the ground plane, and (b) their images, which are established by the currents they induce on the ground plane.

In measuring site attenuation, because of the reflected radiation of the source antenna, there are two waves of electric field strength incident on the receiving antenna from different directions. The incidence angles associated with those directions are shown in Fig. 8-8 as θ_d for the direct wave and θ_r for the reflected wave. Both angles are measured from the vertical direction, which is, of course, normal to the ground plane. The horizontal distance between the source and receiving antennas is assumed to be 10 meters; the source antenna height is fixed at either 1 or 2 meters; and the receiving antenna height is continuously varied from 1 to 4 meters. If h is the horizontal distance between the antennas, s is the source antenna height, and a is the receiving antenna height, then it is seen from Fig. 8-8 that the *direct* distance between the antennas will be

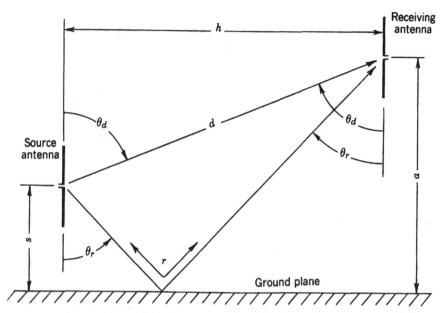

Figure 8-8. Antenna and test-site geometry for the site attenuation measurement.

$$d = \sqrt{h^2 + (a - s)^2} \quad \text{(meters)} \tag{8-52}$$

And the distance traveled by the *reflected* wave, which is the same as the distance from the *image* of the source antenna to the receiving antenna, is

$$r = \sqrt{h^2 + (a + s)^2} \quad \text{(meters)} \tag{8-53}$$

From these equations and Fig. 8-8, it can then be seen that θ_d, the incidence angle of the direct wave, will be such that

$$\sin(\theta_d) = \frac{h}{d} = \frac{h}{\sqrt{h^2 + (a - s)^2}} \tag{8-54}$$

and

$$\cos(\theta_d) = \frac{a - s}{d} = \frac{a - s}{\sqrt{h^2 + (a - s)^2}} \tag{8-55}$$

Thus, when both half-wave dipoles are vertical, the magnitude of their effective lengths relative to the direct wave will be the same and equal to

$$|L_{eff}(\theta_d)| = \frac{\lambda}{\pi} \frac{\cos[(\pi/2)\cos(\theta_d)]}{\sin(\theta_d)} = \frac{\lambda}{\pi} \frac{d}{h} \cos\left[\frac{\pi(a-s)}{2d}\right]$$

$$= \frac{\lambda}{\pi} \frac{\sqrt{h^2 + (a-s)^2}}{h} \cos\left[\frac{\pi(a-s)}{2\sqrt{h^2 + (a-s)^2}}\right] \qquad (8\text{-}56)$$

Also, θ_r, the incidence angle of the reflected wave, will be such that

$$\sin(\theta_r) = \frac{h}{r} = \frac{h}{\sqrt{h^2 + (a+s)^2}} \qquad (8\text{-}57)$$

and

$$\cos(\theta_r) = \frac{a+s}{r} = \frac{a+s}{\sqrt{h^2 + (a+s)^2}} \qquad (8\text{-}58)$$

Therefore, when both half-wave dipoles are vertical, the magnitude of their effective lengths relative to the reflected wave will be the same and equal to

$$|L_{eff}(\theta_r)| = \frac{\lambda}{\pi} \frac{\cos[(\pi/2)\cos(\theta_r)]}{\sin(\theta_r)} = \frac{\lambda}{\pi} \cos\left[\frac{\pi(a+s)}{2r}\right] \frac{r}{h}$$

$$= \frac{\lambda}{\pi} \frac{\sqrt{h^2 + (a+s)^2}}{h} \cos\left[\frac{\pi(a+s)}{2\sqrt{h^2 + (a+s)^2}}\right] \qquad (8\text{-}59)$$

Based on the foregoing observations, it is obvious that variations in the height of *vertical* half-wave dipoles above the ground plane will cause their antenna factors to vary. However, when those dipoles are *horizontal*, they will be parallel to the ground plane, parallel to one another, and perpendicular to the directions of d and r. Thus the effective lengths of both half-wave dipoles will vary only with frequency. In other words, in horizontal polarization, the effective lengths of both dipoles will be $L_{eff}(\theta) = L_{eff}(\pi/2) = \lambda/\pi$.

Variations in the magnitude of the antenna factor when the antenna is a vertical half-wave dipole over the ground plane are shown in the graphs of Fig. 8-9. The plots show variations in the reciprocals of $|L_{eff}(\theta_d)|$ and $|L_{eff}(\theta_r)|$, because, as seen in Eq. 8-41, it is those reciprocals to which the antenna factors are proportional. Thus, the graphs of Fig. 8-9 can be used to correct half-wave dipole antenna factors, as necessary, based on measurement geometry.

In performing site attenuation measurements, the height of the receiving antenna a is varied to obtain the maximum response at each frequency. The relative phases of the direct and reflected waves at incidence depend on $r_\lambda - d_\lambda$, the difference in the distances the waves travel, in wavelengths.

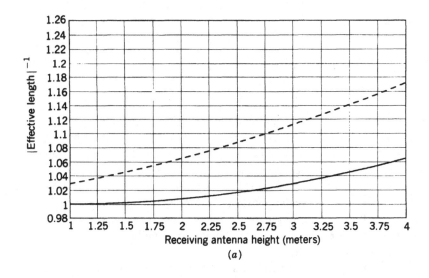

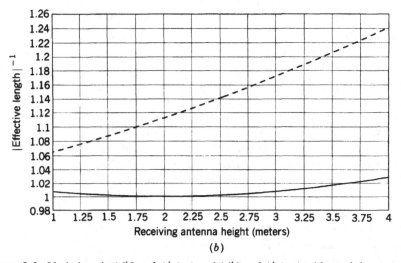

Figure 8-9. Variations in $1/|L_{eff}\theta_d|$ (—) and $1/|L_{eff}\theta_r|$ (---) with receiving antenna height a, when vertical half-wave dipoles are receiving: (a) source height $s = 1$ meter; (b) source height $s = 2$ meters.

Recall that any distance d, in meters, is

$$d_\lambda \equiv \frac{d}{\lambda} = \frac{fd}{c} = \frac{d}{300} f_M \quad \text{(wavelengths)} \quad (8\text{-}60)$$

where f_M is frequency in megahertz, and $c = 300 \times 10^6$ meters/second is the speed of light. From that relationship, together with Eqs. 8-52 and 8-53, it follows that

$$r_\lambda - d_\lambda = \frac{\sqrt{h^2 + (a+s)^2} - \sqrt{h^2 + (a-s)^2}}{300} f_M \quad \text{(wavelengths)} \quad (8\text{-}61)$$

Therefore, based on the phase relationships of antennas and their images, as shown in Fig. 8-7, if the antennas are *horizontal*, then a should be adjusted so as to make $r_\lambda - d_\lambda$ the smallest possible *odd* multiple of half-wavelengths. And, if the antennas are *vertical*, then a should be adjusted so as to make $r_\lambda - d_\lambda$ the smallest possible *even* multiple of half-wavelengths. These adjustments will put the two arriving waves in phase and minimize the distance traveled, thereby maximizing the response of the measuring receiver. The antenna height at which maximum response will occur can be found graphically, as shown in Figs. 8-10 and 8-11, for any given frequency f_M.

To facilitate the measurement of site attenuation, a somewhat more readable graph than those of Figs. 8-10 and 8-11 can be plotted from which to find a_{max} as a function of frequency. These graphs are obtainable by first plotting $r_\lambda - d_\lambda$ for the given value of s, as in Figs. 8-10 and 8-11, then taking a few values of a from those plots and drawing a second graph by interpolation. This graphical procedure is less involved than solving Eq. 8-61 for a_{max}, together with the restriction that $1 \leq a \leq 4$. Typical results are given in Fig. 8-12. These graphs were obtained using Figs. 8-10 and 8-11 to find a few values of a for $s = 2$, then those values were plotted and connected to obtain a_{max} as a function of frequency.

Now, it is clear that any site attenuation measurements made over a ground plane will consist of two components rather than one, as they would in free space. One component results from the electromagnetic wave that travels directly from the source antenna to the receiving antenna; the other component results from the source wave being reflected to the receiving antenna by the ground plane. Thus the voltage V'_R at the measuring receiver will also consist of two components—the voltage V_d caused by the electric field $E_d(t)$ of the *direct* wave and the voltage V_r caused by the electric field $E_r(t)$ of the *reflected* wave.

To fully understand the site attenuation measurement, first suppose that the electric fields of the direct and reflected waves have the same phase when they arrive at the receiving antenna. Also suppose the source and receiving antennas are the same kind of antenna and both have matching baluns. Then, $Z_b = Z_{b'} = Z_a^*$ and from Eqs. 8-45 and 8-48 it can be seen that the current in the transmitting antenna will be

$$|I_0| = \frac{|V_G|}{\sqrt{R_a R_c} K_b \epsilon^{\alpha \ell}}$$

$$= \frac{|V_R| \epsilon^{\alpha \ell} K_b M_{bb'} K_{b'} \epsilon^{\alpha \ell'}}{\sqrt{R_a R_c} K_b \epsilon^{\alpha \ell}} = \frac{|V_R| M_{bb'} K_{b'} \epsilon^{\alpha \ell'}}{\sqrt{R_a R_c}} \quad (8\text{-}62)$$

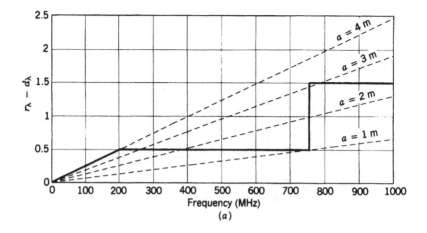

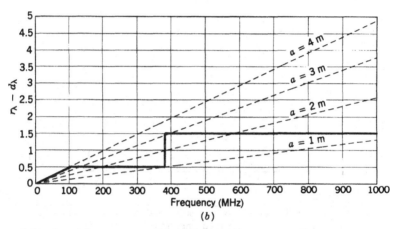

Figure 8-10. Receiving antenna heights of maximum response if $1 \leq a \leq 4$ meters and the antenna is horizontal: (*a*) source height $s = 1$ meter; (*b*) source height $s = 2$ meters.

Given that current in the transmitting antenna, the amplitude of the electric field arriving at the receiving antenna will be

$$|E_R(\theta)| = |E_d(\theta_d) + E_r(\theta_r)|$$

$$= \frac{Z_0|I_0|}{2\lambda} \left[\frac{|L_{eff}(\theta_d)|}{d} + \frac{|L_{eff}(\theta_r)|}{r} \right]$$

$$= \frac{Z_0}{2\lambda} \frac{|V_R|M_{bb'}K_{b'}\epsilon^{\alpha\ell'}}{\sqrt{R_aR_c}} \left[\frac{|L_{eff}(\theta_d)|r + |L_{eff}(\theta_r)|d}{rd} \right]$$

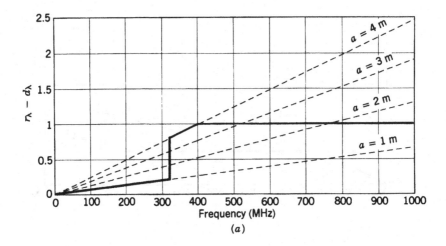

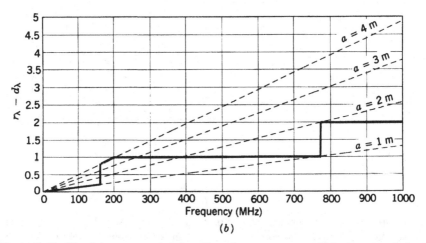

Figure 8-11. Receiving antenna heights of maximum response within the range $1 \leq a \leq$ 4 meters when the antenna is vertical: (*a*) source height $s = 1$ meter; (*b*) source height $s = 2$ meters.

Therefore,

$$\frac{|V_R|}{|E_R(\theta)|} = \frac{2\lambda\sqrt{R_a R_c}}{Z_0 M_{bb'} K_{b'} \epsilon^{\alpha\ell'}} \left[\frac{rd}{|L_{eff}(\theta_d)|r + |L_{eff}(\theta_r)|d} \right] \qquad (8\text{-}63)$$

When $E_d(t)$ and $E_r(t)$ arrive at the receiving antenna with the same phase,

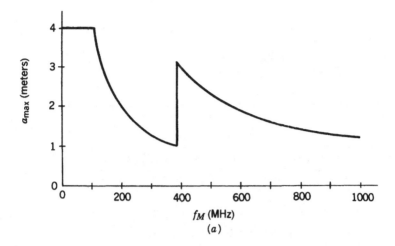

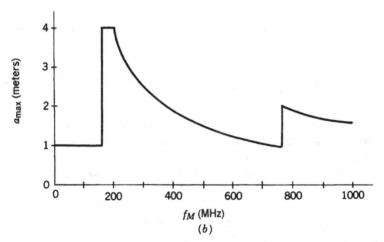

Figure 8-12. Receiving antenna heights at which to expect maximum response with the source antenna 2 meters above the ground plane: (a) horizontal polarization; (b) vertical polarization.

the amplitude of the voltage response at the measuring receiver will be

$$|V_R'| = |V_d + V_r| = \left| \frac{E_d(\theta_d)}{AF(\theta_d)} + \frac{E_r(\theta_r)}{AF(\theta_r)} \right|$$

$$= \left| \frac{AF(\theta_r)E_d(\theta_d) + AF(\theta_d)E_r(\theta_r)}{AF(\theta_d)AF(\theta_r)} \right| \tag{8-64}$$

Thus,

$$
\begin{aligned}
\frac{|E_R(\theta)|}{|V'_R|} &= \left| \frac{AF(\theta_d)AF(\theta_r)[E_d(\theta_d) + E_r(\theta_r)]}{E_d(\theta_d)AF(\theta_r) + E_r(\theta_r)AF(\theta_d)} \right| \\[2mm]
&= \frac{|Z_a + Z_b|}{\sqrt{R_b R_c}} \left| \frac{K_{b'}\epsilon^{\alpha\ell'}L_{eff}^{-1}(\theta_d)L_{eff}^{-1}(\theta_r)[1/d(L_{eff}(\theta_d)) + 1/r(L_{eff}(\theta_r))]}{1/d(L_{eff}(\theta_d)L_{eff}^{-1}(\theta_r)) + 1/r(L_{eff}(\theta_r)L_{eff}^{-1}(\theta_d))} \right| \\[2mm]
&= \frac{2\sqrt{R_a/R_c}K_{b'}\epsilon^{\alpha\ell'}|1/d(L_{eff}^{-1}(\theta_r) + 1/rL_{eff}^{-1}(\theta_d))|}{|1/d(L_{eff}(\theta_d)L_{eff}^{-1}(\theta_r)) + 1/r(L_{eff}(\theta_r)L_{eff}^{-1}(\theta_d))|} \\[2mm]
&= \frac{2[\sqrt{R_a/R_c}K_{b'}\epsilon^{\alpha\ell'}|L_{eff}(\theta_d)r + L_{eff}(\theta_r)d|}{|L_{eff}^2(\theta_d)r + L_{eff}^2(\theta_r)d|}
\end{aligned}
\tag{8-65}
$$

The last of these equations is obtained by multiplying both numerator and denominator by $L_{eff}(\theta_d)L_{eff}(\theta_r)rd$.

Equations 8-63 and 8-65 imply that, when $E_d(t)$ and $E_r(t)$ arrive at the receiving antenna with the same phase, the site attenuation measured will be

$$
SA = \left| \frac{V_R}{V'_R} \right| = \frac{4R_a\lambda}{M_{bb'}Z_0} \left[\frac{rd}{L_{eff}^2(\theta_d)r + L_{eff}^2(\theta_r)d} \right]
\tag{8-66}
$$

This expression is valid for either polarization of the antennas.

Differences in the phases of the electric fields of the direct and reflected waves when they arrive at the receiving antenna can be accounted for as follows. In general, the total electric field arriving at the receiving antenna will be

$$
\begin{aligned}
E_R(t) &= E_d(\theta_d)\cos\left(\omega\left(t - \frac{d}{c}\right)\right) + E_r(\theta_r)\cos\left(\omega\left(t - \frac{r}{c}\right)\right) \\[2mm]
&= \frac{Z_0|I_0|}{2\lambda}\left[\frac{L_{eff}^T(\theta_d)}{d}\cos\left(\omega\left(t - \frac{d}{c}\right)\right) + \frac{L_{eff}^T(\theta_r)}{r}\cos\left(\omega\left(t - \frac{r}{c}\right)\right)\right]
\end{aligned}
\tag{8-67}
$$

When $E_d(t)$ and $E_r(t)$ arrive with the same phase, although $r \neq d$, the total electric field at the receiving antenna will be

$$
\begin{aligned}
E_R(t) &= [E_d(\theta_d) + E_r(\theta_r)]\cos\left(\omega\left(t - \frac{d}{c}\right)\right) \\[2mm]
&= [E_d(\theta_d) + E_r(\theta_r)]\cos\left(\omega\left(t - \frac{r}{c}\right)\right)
\end{aligned}
$$

where

$$E_d(\theta_d) + E_r(\theta_r) = \frac{Z_0|I_0|}{2\lambda} \left[\frac{1}{d} L_{eff}(\theta_d) + \frac{1}{r} L_{eff}(\theta_r) \right] \qquad (8\text{-}68)$$

Now, it was previously assumed in obtaining $|V'_R|$ that the received electric field was that of Eq. 8-68. Therefore, since $|V'_R|$ is the denominator of the site attenuation ratio, it follows that the phase-correction factor will be the ratio of the amplitude $E_R(\theta)$ when $E_d(t)$ and $E_r(t)$ have the same phase, as in Eqs. 8-68, to the amplitude $E_R(\theta)$ when $E_d(t)$ and $E_r(t)$ may be out of phase, as in Eqs. 8-67. That ratio is

$$
\begin{aligned}
\Delta_p &= \frac{|E_d(\theta_d) + E_r(\theta_r)|}{\sqrt{E_d^2(\theta_d) + E_r^2(\theta_r) \pm 2E_d(\theta_d)E_r(\theta_r)\cos(2\pi(r-d)/\lambda)}} \\[2mm]
&= \frac{|rL_{eff}(\theta_d) + dL_{eff}(\theta_r)|}{\sqrt{r^2L_{eff}^2(\theta_d) + d^2L_{eff}^2(\theta_r) \pm 2rdL_{eff}(\theta_d)L_{eff}(\theta_r)\cos(2\pi(r-d)/\lambda)}} \\[2mm]
&\equiv \frac{|rL_{ed} + dL_{er}|}{\sqrt{r^2L_{ed}^2 + d^2L_{er}^2 \pm 2rdL_{ed}L_{er}\cos(2\pi(r-d)/\lambda)}} \qquad (8\text{-}69)
\end{aligned}
$$

The denominator of Δ_p derives from Eqs. 8-67 and the application of identity A-30 of Appendix A with the assumption that $\cos(\omega(t-d/c)) = \cos(x+y)$, and $\cos(\omega(t-r/c)) = \cos(x-y)$. Note that whenever $\omega(t-d/c) = x+y$, and $\omega(t-r/c) = x-y$, it follows that $2y = 2\pi(r-d)/\lambda$. The $\pm$ sign in the denominator of Δ_p is positive for vertically polarized antennas and negative for horizontally polarized antennas, to account for the phase reversal when horizontal electric fields are reflected by the ground plane.

Thus, for any given antenna heights, the site attenuation measured on an ideal test site with like vertical antennas that have matching baluns will be

$$SA_V = \frac{4R_a\lambda}{M_{bb'}Z_0} \frac{rd[rL_{ed} + dL_{er}][L_{er}^2 d + L_{ed}^2 r]^{-1}}{\sqrt{r^2L_{ed}^2 + d^2L_{er}^2 + 2rdL_{ed}L_{er}\cos(2\pi(r-d)/\lambda)}} \qquad (8\text{-}70)$$

and with like horizontal antennas site attenuation will be

$$SA_H = \frac{4R_a\lambda}{M_{bb'}Z_0} \frac{rd[rL_{ed} + dL_{er}][L_{er}^2 d + L_{ed}^2 r]^{-1}}{\sqrt{r^2L_{ed}^2 + d^2L_{er}^2 - 2rdL_{ed}L_{er}\cos(2\pi(r-d)/\lambda)}} \qquad (8\text{-}71)$$

For example, if the antennas are both *vertical half-wave dipoles* with matching baluns

$$L_{ed} = \frac{\lambda}{\pi} \frac{\cos(\pi/2(\cos \theta_d))}{\sin \theta_d} = \frac{\lambda d}{\pi h} \cos\left(\frac{\pi}{2} \cos \theta_d\right) \equiv \frac{\lambda d}{\pi h} C_d$$

and

$$L_{er} = \frac{\lambda}{\pi} \frac{\cos(\pi/2(\cos \theta_r))}{\sin \theta_r} = \frac{\lambda r}{\pi h} \cos\left(\frac{\pi}{2} \cos \theta_r\right) \equiv \frac{\lambda r}{\pi h} C_r \qquad (8\text{-}72)$$

Therefore, Eq. 8-70 becomes

$$SA_V = \frac{4R_D\lambda}{M_{bb'}Z_0} \frac{\pi^2 h^2}{\lambda^2} \frac{[C_d + C_r][rC_r^2 + dC_d^2]^{-1}}{\sqrt{C_d^2 + C_r^2 + 2C_dC_r \cos(2\pi(r-d)/\lambda)}}$$

$$= 2.13 f_M \frac{[C_d + C_r][rC_r^2 + dC_d^2]^{-1}}{\sqrt{C_d^2 + C_r^2 + 2C_dC_r \cos(2\pi(r-d)/\lambda)}} \qquad (8\text{-}73)$$

The coefficient of the latter expression results because $R_D = 73.2$ ohms, $h = 10$ meters, $M_{bb'} = |2Z_D^*|/(2R_D) = 1.2$, $Z_0 = 120\pi$ ohms, and $\lambda = 300 \times 10^6$ m/sec.

When the transmitting and receiving antennas are both *horizontal half-wave dipoles* with matching baluns, then

$$L_{ed} = L_{er} = \frac{\lambda}{\pi} \qquad (8\text{-}74)$$

and Eq. 8-71 becomes

$$SA_H = \frac{4R_a\lambda}{M_{bb'}Z_0} \frac{\pi^2}{\lambda^2} \frac{rd}{\sqrt{r^2 + d^2 - 2rd\cos(2\pi(r-d)/\lambda)}}$$

$$= \frac{2.13 f_M}{100} \frac{rd}{\sqrt{r^2 + d^2 - 2rd\cos(2\pi(r-d)/\lambda)}} \qquad (8\text{-}75)$$

Sample values of 20 log(SA_V) and 20 log(SA_H)—site attenuation in decibels (dB)—obtained with Eqs. 8-73 and 8-75, and receiving antenna heights obtained from Figs. 8-10 and 8-11, are given in Tables 8-1 and 8-2.

TABLE 8-1. Site Attenuation for Vertical Half-Wave Dipoles When
h = 10 meters and s = 2 meters

	f_M (MHz)									
	100	200	300	400	500	600	700	800	900	1000
a (meters)	1.0	4.0	2.6	1.9	1.6	1.3	1.1	1.9	1.7	1.6
SA$_V$ (dB)	22.1	29.2	31.6	33.7	35.6	37.0	38.3	39.8	40.7	41.7

8.4.2.3 Normalized Site Attenuation To obtain a standard for test-site validation that is independent of the antennas and the measurement systems used, a normalized expression for site attenuation can be defined. *Normalized site attenuation*, or NSA, is nothing more than the power transmission loss that would occur on the ideal test site between nondirectional, or isotropic, transmitting and receiving antennas.

Thus, to calculate NSA it is assumed that $AF_T(\theta) = AF_R(\theta) = 1$ m^{-1}, which implies that the antennas are nondirectional, and that there are no dissipative losses of any kind in the measurement system. Based on those assumptions,

$$AF_T(\theta) = AF_R(\theta) = \frac{|Z_a + Z_b|}{\sqrt{R_b R_c} L_{eff}(\theta)} = 2\sqrt{\frac{R_a}{R_c}} \frac{1}{L_{eff}(\theta)} = 1$$

and

$$L_{eff}(\theta) = 2\sqrt{\frac{R_a}{R_c}} = L_{ed} = L_{er} \tag{8-76}$$

Therefore, Eqs. 8-70 and 8-71 become

TABLE 8-2. Site Attenuation for Horizontal Half-Wave Dipoles When
h = 10 meters and s = 2 meters

	f_M (MHz)									
	100	200	300	400	500	600	700	800	900	1000
a (meters)	4.0	2.0	1.3	3.0	2.3	1.8	1.6	1.4	1.3	1.2
SA$_H$(dB)	21.3	26.9	30.3	33.1	34.9	36.5	37.8	39.0	39.9	40.9

$$\text{NSA}_V = \frac{R_c\lambda}{Z_0} \frac{rd}{\sqrt{r^2 + d^2 + 2rd\cos(2\pi(r-d)/\lambda)}}$$

$$= \frac{125}{\pi f_M} \frac{rd}{\sqrt{r^2 + d^2 + 2rd\cos(2\pi(r-d)/\lambda)}} \qquad (8\text{-}77)$$

and

$$\text{NSA}_H = \frac{R_c\lambda}{Z_0} \frac{rd}{\sqrt{r^2 + d^2 - 2rd\cos(2\pi(r-d)/\lambda)}}$$

$$= \frac{125}{\pi f_M} \frac{rd}{\sqrt{r^2 + d^2 - 2rd\cos(2\pi(r-d)/\lambda)}} \qquad (8\text{-}78)$$

In obtaining the latter of both Eqs. 8-77 and 8-78 it is assumed that $R_c = 50$ ohms.

Tables 8-3 and 8-4 give sample values of NSA_H and NSA_V in decibels (dB) obtained with these equations and antenna heights from Figs. 8-10 and 8-11.

8.4.2.4 Environment, Geometry, and Antenna Impedance Another effect of significance in the site attenuation measurement is the effect of the ground plane on the impedance of the half-wave dipole. Impedance changes also cause changes in the antenna factors associated with a half-wave dipole as the measurement geometry is changed. This was mentioned above, and is now examined in more detail.

Although it has not been specifically discussed until now, it should be clear that when any conductor has a current induced in or on it by electromagnetic radiation, the induced current will also radiate. In other words, time-varying electric currents radiate electromagnetic energy, regardless of their origin. Thus, over a ground plane, both source antennas and receiving antennas will be affected by currents induced in them by their own images. They will radiate, establishing those images, they will receive reflected radiations back from their images, the currents induced by the reflected radiations will reradiate, and so

TABLE 8-3. Normalized Site Attenuation for Horizontal Polarization When $h = 10$ meters and $s = 1$ meter

	f_M (MHz)									
	100	200	300	400	500	600	700	800	900	1000
a (meters)	4	4	2.6	1.9	1.5	1.3	1.1	2.9	2.6	2.3
NSA_H (dB)	9.7	0.6	−3.3	−5.9	−7.9	−9.5	−10.8	−11.7	−12.8	−13.8

TABLE 8-4. Normalized Site Attenuation for Vertical Polarization When
$h = 10$ **meters and** $s = 2.75$ **meter**

	f_M (MHz)									
	100	200	300	400	500	600	700	800	900	1000
a (meters)	1	3	1.9	1.4	1.2	1.9	1.6	1.4	1.3	1.1
NSA$_V$ (dB)	7.7	0.6	−3.1	−5.7	−7.5	−9.1	−10.5	−11.7	−12.6	−13.5

on, until the total distance traveled back and forth causes the radiations and reradiations to be insignificant. The effect this has is to establish currents in the antennas that are not equal to the applied voltages divided by the free-space impedance of the antennas. In other words, because of the presence of the ground plane, antenna impedances may change.

This effect of the ground plane on the antennas is quite involved to analyze mathematically. However, it has been done, and the results are given in many of the antenna engineering books listed in the bibliography. The values of the mutual resistance and mutual reactance that are induced in a half-wave dipole by its image in the ground plane are shown graphically here. They are given in Fig. 8-13 for a horizontal half-wave dipole and in Fig. 8-14 for a vertical half-wave dipole. These values of mutual impedance subtract from Z_D to give the total impedance for horizontal dipoles, and they add to Z_D to give the total impedance for vertical dipoles. This is due to the phase relationships of the currents in the dipoles and their images, as shown in Fig. 8-7b. Thus, for a horizontal half-wave dipole above a ground plane, $Z_D = (73.2 - R_m) + j(42.5 - X_m)$, and for a vertical half-wave dipole over a ground plane, $Z_D = (73.2 + R_m) + j(42.5 + X_m)$.

The effect these changes have on site attenuation is caused by changes in the value of $|Z_D + Z_b|$ in both antenna factors. Those changes are shown graphically in Fig. 8-13 for horizontal half-wave dipoles and in Fig. 8-14 for vertical half-wave dipoles. It is seen from those graphs that the changes in this case are more significant for horizontal dipoles than they are for vertical dipoles. In either case, the closer the dipole is to the ground plane, the greater the change in its impedance will be.

Changes in site attenuation caused by changes in half-wave dipole impedances can be evaluated as follows. If Z'_D and Z''_D are the impedances of the transmitting and receiving dipoles, then $|Z_a + Z_b|^2$ should be replaced by $|Z'_D + Z_b||Z''_D + Z_b|$ to obtain appropriate expressions for site attenuation. Since site attenuation is proportional to one of these quantities, any change in site attenuation due to changes in dipole impedances will equal the ratio of $|Z'_D + Z_b||Z''_D + Z_b|$ to $|Z_D + Z_b|^2$. For example, referring to Table 8-2, it is seen that at $f_M = 100$ MHz, $a_\lambda = 1.33$; thus, from Fig. 8-14, it can be seen that $|Z'_D + Z_b| = 153$ ohms. Also, $s_\lambda = 2f_M/300 = 0.67$, so that $|Z''_D + Z_b| = 136$

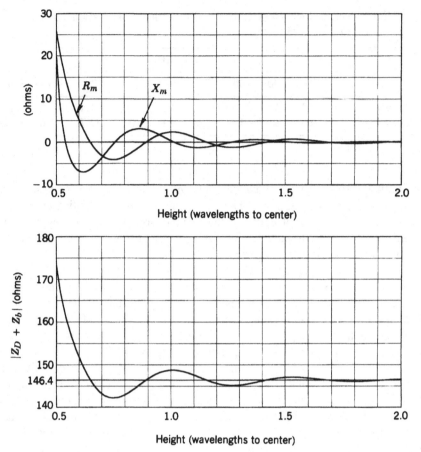

Figure 8-13. Impedance changes (R_m, X_m) in a vertical half-wave dipole, caused by its ground plane image, and their effect on the value of $|Z_D + Z_b|$.

ohms. Therefore, the change ratio will be

$$\frac{|Z'_D + Z_b||Z''_D + Z_b|}{|Z_D + Z_b|^2} = \frac{153(136)}{(146.4)^2} = 0.97 \tag{8-79a}$$

And at $f_M = 200$ MHz, $a_\lambda = 1.33$, so that $|Z'_D + Z_b| = 153$ ohms, and $s_\lambda = 1.33$, so that $|Z''_D + Z_b| = 153$ ohms. Therefore, in this case,

$$\frac{|Z'_D + Z_b||Z''_D + Z_b|}{|Z_D + Z_b|^2} = \frac{153^2}{(146.4)^2} = 1.09 \tag{8-79b}$$

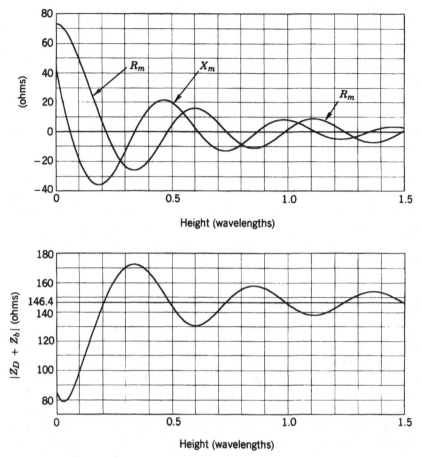

Figure 8-14. Impedance changes (R_m, X_m) in a horizontal half-wave dipole, caused by its ground plane image, and their effect on the value of $|Z_D + Z_b|$.

Thus, in the first example, the change in site attenuation will be 20 log(0.97) = −0.26 dB, and in the second example, the change will be 20 log(1.09) = 0.77 dB. Neither of these examples indicate that the changes caused will be terribly significant. Nevertheless, the potential for error caused by these impedance changes should not be overlooked.

One last observation should be made regarding test-site validation. It is clear that a fixed antenna geometry would yield the same values of site attenuation at any location on the ideal test site. Therefore, the same should be true for any actual test site that is a valid simulation of the ideal. In other words, at any location on a test site, at any given frequency, for fixed values of h, a, and s, the measured site attenuation should be exactly the same. Measurements made with this in mind are very useful for troubleshooting, problem isolation, and gaining general assurance that a test site is behaving properly.

8.5 ROUTINE RADIATION MEASUREMENTS

After accurate antenna factors have been established for a measurement system and a test site has been validated, the routine measurement of electromagnetic radiations is relatively straightforward. Many of the results obtained and procedures developed while performing site attenuation measurements are also useful in preparing for and making measurements to find and reduce unintentional radiations. Nevertheless, properly performed measurements of unintentional electromagnetic radiations will often be time-consuming.

Equipment being tested must be examined from all directions, because its radiation patterns will vary in magnitude and shape with direction and frequency. Therefore, receiving antenna height must be varied over the full range at each frequency. Also, the equipment being tested must be fully rotated so that it can be observed from every angle at each frequency. This can be facilitated somewhat by using broadband antennas rather than half-wave dipoles to find and reduce unintentional radiations.

8.5.1 Broadband Antenna Calibration

In making routine radiation measurements, it is considerably less time-consuming and much more convenient to use broadband antennas rather than dipoles, because the former do not require physical tuning at each measurement frequency. Of course, any antennas used must be accurately calibrated. That can best be done by comparing their responses to those of previously calibrated half-wave dipoles.

Broadband antenna calibration with a half-wave dipole having a known antenna factor is quite straightforward. Both are used to measure the same radiation. At each measurement frequency, the difference in their responses is added to the antenna factor for the half-wave dipole to obtain the antenna factor for the broadband antenna at that frequency. The new antenna factors should be determined by measurement at as many frequencies as possible, then their values at the remaining frequencies, where measurements were not made, can be approximated by linear interpolation. However, changes in antenna factors from one frequency to the next are not always linear. Therefore, measurements should always be made to verify the linear interpolations if any questions or inconsistencies should arise.

Broadband antennas can be calibrated by using three different broadband antennas in three different pairs to measure site attenuation on a valid test site. For this purpose, site attenuation is usually defined to include balun losses. Thus, the measurement before insertion is made by simply connecting the two antenna cables directly together. So, for antenna calibration, site attenuation is defined as

$$SA' \equiv K_b M_{bb'} K_{b'} SA \Delta_p$$

$$= \frac{4R_a \lambda}{Z_0} \left[\frac{K_b K_{b'} rd}{L_{eff}^2(\theta_d)r + L_{eff}^2(\theta_r)d} \right] \Delta_p \qquad (8\text{-}80)$$

The antenna factors found in this way do not include cable loss.

To obtain a known quantity equal only to the losses in a given pair of antennas and their baluns, measured site attenuation (SA′) is divided by the normalized site attenuation (NSA). Three identical site attenuation measurements are made for the geometries of interest, in the frequency range of interest, using three different pairs of three antennas. Three equations with three unknowns, which are the antenna factors sought result. The measured results are

$$AF'_1(\theta)AF'_2(\theta) = \frac{SA'_1}{NSA}$$

$$AF'_2(\theta)AF'_3(\theta) = \frac{SA'_2}{NSA}$$

and

$$AF'_3(\theta)AF'_1(\theta) = \frac{SA'_3}{NSA} \tag{8-81}$$

From these equations it is seen that

$$AF'_1(\theta) = \sqrt{\frac{SA'_1 SA'_3}{SA'_2 NSA}}$$

$$AF'_2(\theta) = \sqrt{\frac{SA'_1 SA'_2}{SA'_3 NSA}}$$

and

$$AF'_3(\theta) = \sqrt{\frac{SA'_2 SA'_3}{SA'_1 NSA}} \tag{8-82}$$

The antenna factors thus obtained will, of course, include the effects of both the direct and the reflected waves, and they will be dependent upon the measurement geometry at any given frequency. Therefore, whenever they are used at each frequency they must be associated with the same values of h, s, and a as those at which the site attenuation measurements were made.

Thus, given a validated test site and calibrated receiving antennas, the measurement of unnecessary electromagnetic radiations can be performed with relative ease. The measurements will be time-consuming, but the steps of their performance will be straightforward. Quite simply, any equipment under test must be viewed from all directions leading away from it: once with the antennas horizontally polarized and once with the antennas vertically polarized. At each direction from the equipment, the antennas must be raised and lowered over the full height range prescribed at each measurement frequency. Finally, the maximum field strength received by the antenna must be recorded for each frequency and polarization.

8.6 SUMMARY

The test sites on which the measurement of unintentional electromagnetic radiations are performed are simulations of an infinite, empty half-space, bounded by an infinite, perfectly conducting ground plane. These simulations of an empty half-space often consist of large, flat, open areas—large fields, large flat rooftops, or empty parking lots—the surfaces of which are covered with continuous metal sheeting or screening that is well grounded. The ideal measurement environment can also be simulated by a metal chamber in which the inner walls and ceiling are covered with electromagnetically anechoic material, and the metal floors are highly conductive and well grounded.

The basic measurement system with which the measurements of unintentional electromagnetic radiations are performed consists of a well-calibrated receiving antenna that is connected by coaxial cable to a measuring receiver or spectrum analyzer. The antenna is positioned above the ground plane, and the measuring receiver is somehow located outside the simulated empty half-space. The measuring receiver and its observers are usually located beneath the ground plane of an outdoor test site, or outside the enclosure of a partially anechoic chamber. A typical measurement setup is that illustrated in Fig. 1-2 of Chapter 1.

The site attenuation measurement is performed to insure that actual test sites provide valid simulations of the ideal test site. The apparent emptiness of the half-space above the test site and the near-total reflectivity of electromagnetic radiation by its ground plane are the attributes that must be verified. Given satisfactory site attenuation measurements and properly calibrated receiving antennas, the day-to-day measurement of unintentional electromagnetic radiations can then be performed with accuracy and little difficulty.

APPENDIX A

USEFUL TRIGONOMETRIC IDENTITIES

A.1 BASIC IDENTITIES

A number of basic trigonometric identities that have gone into the development of the discussions in this book are presented here.

For any real numbers x and y,

$$\sin(x + y) = \sin(x)\cos(y) + \sin(y)\cos(x) \qquad \text{(A-1)}$$

$$\sin(x - y) = \sin(x)\cos(y) - \sin(y)\cos(x) \qquad \text{(A-2)}$$

$$\cos(x + y) = \cos(x)\cos(y) - \sin(x)\sin(y) \qquad \text{(A-3)}$$

$$\cos(x - y) = \cos(x)\cos(y) + \sin(x)\sin(y) \qquad \text{(A-4)}$$

By combining pairs of the above relationships, the following identities are obtained.

$$\sin(x + y) + \sin(x - y) = 2\sin(x)\cos(y) \qquad \text{(A-5)}$$

$$\sin(x + y) - \sin(x - y) = 2\cos(x)\sin(y) \qquad \text{(A-6)}$$

$$\cos(x + y) + \cos(x - y) = 2\cos(x)\cos(y) \qquad \text{(A-7)}$$

$$\cos(x - y) - \cos(x + y) = 2\sin(x)\sin(y) \qquad \text{(A-8)}$$

And, from these relationships, recalling that $\sin(0) = 0$ and $\cos(0) = 1$, if $y = x$, then

$$\sin(2x) = 2\sin(x)\cos(x) \tag{A-9}$$

$$\sin^2(x) + \cos^2(x) = 1 \tag{A-10}$$

and

$$\cos(2x) = \cos^2(x) - \sin^2(x) \tag{A-11}$$

Then, from Eqs. A-9, A-10, and A-11, it also follows that

$$1 + \cos(2x) = 2\cos^2(x) \tag{A-12}$$
$$1 - \cos(2x) = 2\sin^2(x) \tag{A-13}$$
$$1 + \sin(2x) = [\sin(x) + \cos(x)]^2 \tag{A-14}$$

and

$$1 - \sin(2x) = [\sin(x) - \cos(x)]^2 \tag{A-15}$$

Also,

$$
\begin{aligned}
\sin(x+y)&\sin(x-y)\\
&= [\sin(x)\cos(y) + \sin(y)\cos(x)][\sin(x)\cos(y) - \sin(y)\cos(x)]\\
&= [\sin(x)\cos(y)]^2 - [\sin(y)\cos(x)]^2\\
&= \sin^2(x)[1 - \sin^2(y)] - \sin^2(y)\cos^2(x)\\
&= \sin^2(x) - \sin^2(y)[\sin^2(x) + \cos^2(x)]\\
&= \sin^2(x) - \sin^2(y)
\end{aligned}
\tag{A-16}
$$

And

$$
\begin{aligned}
\cos(x+y)&\cos(x-y)\\
&= [\cos(x)\cos(y) - \sin(x)\sin y)][\cos(x)\cos(y) + \sin(x)\sin(y)]\\
&= [\cos(x)\cos(y)]^2 - [\sin(x)\sin(y)]^2\\
&= \cos^2(x)[1 - \sin^2(y)] - \sin^2(y)\sin^2(x)\\
&= \cos^2(x) - \sin^2(y)[\cos^2(x) + \sin^2(x)]\\
&= \cos^2(x) - \sin^2(y)
\end{aligned}
\tag{A-17}
$$

From Eqs. A-2 and A-4, if $x = 0$, then

$$\sin(-y) = -\sin(y) \tag{A-18}$$

and

$$\cos(-y) = \cos(y) \tag{A-19}$$

From Eqs. A-1, A-2, A-3, and A-4, if $y = \pi/2$, then

$$\sin\left(x + \frac{\pi}{2}\right) = \cos(x) \tag{A-20}$$

$$\sin\left(x - \frac{\pi}{2}\right) = -\cos(x) \tag{A-21}$$

$$\cos\left(x + \frac{\pi}{2}\right) = -\sin(x) \tag{A-22}$$

and

$$\cos\left(x - \frac{\pi}{2}\right) = \sin(x) \tag{A-23}$$

And, also from Eqs. A-1, A-2, A-3, and A-4, if $y = \pi$, then

$$\sin(x \pm \pi) = -\sin(x) \tag{A-24}$$

and

$$\cos(x \pm \pi) = -\cos(x) \tag{A-25}$$

A.2   FURTHER RELATIONSHIPS

A number of further relationships based on the above identities were found to be particularly useful. Whenever $\zeta = \arctan(B/A)$, then $\sin(\zeta) = B/\sqrt{A^2 + B^2}$ and $\cos(\zeta) = A/\sqrt{A^2 + B^2}$. Therefore,

$$
\begin{aligned}
A\sin(x) + B\cos(x) &= \sqrt{A^2 + B^2}\left[\frac{A\sin(x)}{\sqrt{A^2 + B^2}} + \frac{B\cos(x)}{\sqrt{A^2 + B^2}}\right] \\
&= \sqrt{A^2 + B^2}[\sin(x)\cos(\zeta) + \sin(\zeta)\cos(x)] \\
&= \sqrt{A^2 + B^2}\sin(x + \zeta)
\end{aligned}
$$

In other words, if $\zeta = \arctan(B/A)$, then

$$A \sin(x) + B \cos(x) = \sqrt{A^2 + B^2} \sin(x + \zeta) \tag{A-26}$$

Similarly

$$A \sin(x) - B \cos(x) = \sqrt{A^2 + B^2} \sin(x - \zeta) \tag{A-27}$$

$$A \cos(x) + B \sin(x) = \sqrt{A^2 + B^2} \cos(x - \zeta) \tag{A-28}$$

and

$$A \cos(x) - B \sin(x) = \sqrt{A^2 + B^2} \cos(x + \zeta) \tag{A-29}$$

Also, since

$$A \cos(x + y) = A[\cos(x) \cos(y) - \sin(x) \sin(y)]$$

and

$$B \cos(x - y) = B[\cos(x) \cos(y) + \sin(x) \sin(y)]$$

it follows that

$$
\begin{aligned}
A \cos(x + y) + B \cos(x - y) &= (A + B) \cos(x) \cos(y) - (A - B) \sin(x) \sin(y) \\
&= [(A + B) \cos(y)] \cos(x) - [(A - B) \sin(y)] \sin(x) \\
&= \sqrt{(A + B)^2 \cos^2(y) + (A - B)^2 \sin^2(y)} \cos(x + \zeta) \\
&= \sqrt{A^2 + B^2 + 2AB(\cos^2(y) - \sin^2(y))} \cos(x + \zeta) \\
&= \sqrt{A^2 + 2AB \cos(2y) + B^2} \cos(x + \zeta)
\end{aligned}
$$

In summary, then

$$A \cos(x + y) + B \cos(x - y) = \sqrt{A^2 + 2AB \cos(2y) + B^2} \cos(x + \zeta) \tag{A-30}$$

where

$$\zeta = \arctan\left[\frac{(A - B) \sin(y)}{(A + B) \cos(y)}\right] = \arctan\left[\frac{(A - B) \tan(y)}{A + B}\right]$$

And, similarly,

$$A \cos(x + y) - B \cos(x - y)$$
$$= (A - B) \cos(x) \cos(y) - (A + B) \sin(x) \sin(y)$$
$$= [(A - B) \cos(y)] \cos(x) - [(A + B) \sin(y)] \sin(x)$$
$$= \sqrt{(A - B)^2 \cos^2(y) + (A + B)^2 \sin^2(y)} \cos(x + \zeta)$$
$$= \sqrt{A^2 + B^2 - 2AB(\cos^2(y) - \sin^2(y))} \cos(x + \zeta)$$
$$\sqrt{A^2 - 2AB \cos(2y) + B^2} \cos(x + \zeta)$$

so that

$$A \cos(x + y) - B \cos(x - y) = \sqrt{A^2 - 2AB \cos(2y) + B^2} \cos(x + \zeta) \qquad \text{(A-31)}$$

where

$$\zeta = \arctan\left[\frac{(A + B) \sin(y)}{(A - B) \cos(y)}\right] = \arctan\left[\frac{(A + B) \tan(y)}{A - B}\right]$$

Using similar reasoning it is seen that

$$A \sin(x + y) + B \sin(x - y) = \sqrt{A^2 - 2AB \cos(2y) + B^2} \sin(x + \zeta) \qquad \text{(A-32)}$$

where

$$\zeta = \arctan t\left[\frac{(A + B) \cos(y)}{(A - B) \sin(y)}\right] = \arctan\left[\frac{A + B}{(A - B) \tan(y)}\right]$$

And,

$$A \sin(x + y) - B \sin(x - y) = \sqrt{A^2 + 2AB \cos(2y) + B^2} \sin(x + \zeta) \qquad \text{(A-33)}$$

where

$$\zeta = \arctan\left[\frac{(A - B) \cos(y)}{(A + B) \sin(y)}\right] = \arctan\left[\frac{A - B}{(A + B) \tan(y)}\right]$$

And, finally, it should be noted that, for any x, A, and B,

$$|A \sin(x) + B \cos(x)| = \sqrt{A^2 + B^2} \left| \frac{A \sin(x)}{\sqrt{A^2 + B^2}} + \frac{B \cos(x)}{\sqrt{A^2 + B^2}} \right|$$

$$= \sqrt{A^2 + B^2} \, | \sin(x) \cos(y) + \sin(y) \cos(x)|$$

$$= \sqrt{A^2 + B^2} \, | \sin(x + y)|$$

$$\leq \sqrt{A^2 + B^2} \tag{A-34}$$

A.3 RADIATION PATTERN FACTOR EXPRESSIONS

The pattern factor $E_T(\theta)$ given in Eq. 2-24 can be evaluated when L_λ is an integer multiple of half wavelengths as follows. Given that

$$E_L(\theta) = \frac{Z_0 |I|}{2\pi d} \sin \theta \, \frac{\sin[\pi L_\lambda (\cos \theta - 1)]}{\cos \theta - 1}$$

and

$$E_L(\theta - \pi) = -\frac{Z_0 |I|}{2\pi d} \sin \theta \, \frac{\sin[\pi L_\lambda (\cos \theta + 1)]}{\cos \theta + 1}$$

if $L_\lambda = n$, where $n = 1, 2, 3, \ldots,$ any positive integer, then

$$\cos(2\pi L_\lambda) = \cos(2\pi n) = 1$$

$$\sin(\pi L_\lambda) = 0$$

$$\cos(\pi L_\lambda) = 1 \text{ when } n = 2, 4, 6, \ldots$$

$$\cos(\pi L_\lambda) = -1 \text{ when } n = 1, 3, 5, \ldots$$

$$\sin(\pi L_\lambda \cos \theta \pm \pi L_\lambda) = \pm \cos(\pi L_\lambda) \sin(\pi L_\lambda \cos \theta)$$

And,

$$E_T(\theta) = E_L(\theta) + E_L(\theta - \pi)$$

$$= \frac{Z_0 |I| \sin \theta}{2\pi d} \left[\frac{\sin(\pi L_\lambda \cos \theta - \pi L_\lambda)}{\cos \theta - 1} - \frac{\sin(\pi L_\lambda \cos \theta + \pi L_\lambda)}{\cos \theta + 1} \right]$$

$$= \frac{Z_0 |I| \sin \theta}{2\pi d} \left[\frac{\mp \sin(\pi L_\lambda \cos \theta)}{\cos \theta - 1} \mp \frac{\sin(\pi L_\lambda \cos \theta)}{\cos \theta + 1} \right]$$

$$= \frac{Z_0|I|\sin\theta}{2\pi d}\left[\frac{\mp 2\sin(\pi L_\lambda\cos\theta)}{\cos^2\theta - 1}\right]$$

$$= \pm\frac{Z_0|I|}{\pi d}\frac{\sin(\pi L_\lambda\cos\theta)}{\sin\theta} \tag{A-35}$$

This expression for $E_T(\theta)$ is positive when $n = 1, 3, 5, \ldots$, and negative when $n = 2, 4, 6, \ldots$.

Similarly, if $L_\lambda = n/2$, where $n = 1, 3, 5, \ldots$, an odd integer, then

$$\cos(2\pi L_\lambda) = \cos(n\pi) = -1$$
$$\cos(\pi L_\lambda) = 0$$
$$\sin(\pi L_\lambda) = +1 \text{ when } n = 1, 5, 9, \ldots$$
$$\sin(\pi L_\lambda) = -1 \text{ when } n = 3, 7, 11, \ldots$$
$$\sin(\pi L_\lambda\cos\theta \pm \pi L_\lambda) = \pm\sin(\pi L_\lambda)\cos(\pi L_\lambda\cos\theta)$$

And,

$$E_T(\theta) = E_L(\theta) - E_L(\theta - \pi)$$

$$= \frac{Z_0|I|\sin\theta}{2\pi d}\left[\frac{\sin(\pi L_\lambda\cos\theta - \pi L_\lambda)}{\cos\theta - 1} + \frac{\sin(\pi L_\lambda\cos\theta + \pi L_\lambda)}{\cos\theta + 1}\right]$$

$$= \frac{Z_0|I|\sin\theta}{2\pi d}\left[\frac{\mp\cos(\pi L_\lambda\cos\theta)}{\cos\theta - 1} \pm \frac{\cos(\pi L_\lambda\cos\theta)}{\cos\theta + 1}\right]$$

$$= \frac{Z_0|I|\sin\theta}{2\pi d}\left[\frac{\mp 2\cos(\pi L_\lambda\cos\theta)}{\cos^2\theta - 1}\right]$$

$$= \pm\frac{Z_0|I|}{\pi d}\frac{\cos(\pi L_\lambda\cos\theta)}{\sin\theta} \tag{A-36}$$

The resulting expression for $E_T(\theta)$ is positive when $n = 3, 7, 11, \ldots$, and negative when $n = 1, 5, 9, \ldots$.

It follows from all of the above that when L_λ equals an *even* number of half wavelengths,

$$|E_T(\theta)| = \left|\frac{Z_0|I|}{\pi d}\frac{\sin(\pi L_\lambda\cos\theta)}{\sin\theta}\right|$$

and when L_λ equals an *odd* number of half wavelengths,

$$|E_T(\theta)| = \left| \frac{Z_0|I|}{\pi d} \frac{\cos(\pi L_\lambda \cos \theta)}{\sin \theta} \right|$$

A.4   TRANSITION FACTOR EXPRESSIONS

Each of the transition factors D_n in Chapter 4 is the square root of a sum of two squared expressions. The first of those expressions has the general form

$$\left[\frac{\sin^2(x)}{t_1} - \frac{\sin(y)\sin(z)}{t_2} \right]^2$$

$$= \frac{\sin^4(x)}{t_1^2} - 2 \frac{\sin^2(x)}{t_1} \frac{\sin(y)\sin(z)}{t_2} + \frac{\sin^2(y)\sin^2(z)}{t_2^2}$$

$$= \frac{\sin^2(x)\sin^2(x)}{t_1^2} - 2 \frac{\sin(x)\sin(y)}{t_1 t_2} \sin(x)\sin(z) + \frac{\sin^2(y)\sin^2(z)}{t_2^2}$$

And, the second has the general form

$$\left[\frac{\sin(x)\cos(x)}{t_1} - \frac{\sin(y)\cos(z)}{t_2} \right]^2$$

$$= \frac{\sin^2(x)\cos^2(x)}{t_1^2} - 2 \frac{\sin(x)\cos(x)}{t_1} \frac{\sin(y)\cos(z)}{t_2} + \frac{\sin^2(y)\cos^2(z)}{t_2^2}$$

$$= \frac{\sin^2(x)\cos^2(x)}{t_1^2} - 2 \frac{\sin(x)\sin(y)}{t_1 t_2} \cos(x)\cos(z) + \frac{\sin^2(y)\cos^2(z)}{t_2^2}$$

Therefore, applying the identities A-10, and A-4, given above, the general form of the sum of these two quantities is seen to be

$$\left[\frac{\sin^2(x)}{t_1} - \frac{\sin(y)\sin(z)}{t_2} \right]^2 + \left[\frac{\sin(x)\cos(x)}{t_1} - \frac{\sin(y)\cos(z)}{t_2} \right]^2$$

$$= \frac{\sin^2(x)[\sin^2(x) + \cos^2(x)]}{t_1^2}$$

$$- 2 \frac{\sin(x)\sin(y)}{t_1 t_2} [\cos(x)\cos(z) + \sin(x)\sin(z)]$$

$$+ \frac{\sin^2(y)[\sin^2(z) + \cos^2(z)]}{t_2^2}$$

$$= \frac{\sin^2(x)}{t_1^2} - 2\,\frac{\sin(x)\sin(y)}{t_1 t_2}\,\cos(z - x) + \frac{\sin^2(y)}{t_2^2} \qquad \text{(A-37)}$$

Thus, the general form of the transition factors defined in Chapter 4 is

$$D_n = \sqrt{\left(\frac{\sin^2(x)}{t_1} - \frac{\sin(y)\sin(z)}{t_2}\right)^2 + \left(\frac{\sin(x)\cos(x)}{t_1} - \frac{\sin(y)\cos(z)}{t_2}\right)^2}$$

$$= \sqrt{\frac{\sin^2(x)}{t_1^2} - 2\,\frac{\sin(x)\sin(y)}{t_1 t_2}\,\cos(z - x) + \frac{\sin^2(y)}{t_2^2}} \qquad \text{(A-38)}$$

APPENDIX B

BASIC VOLTAGE WAVEFORM FOURIER COEFFICIENTS

B.1 INDEFINITE INTEGRALS

The following indefinite integrals will be used in determining the Fourier coefficients of the basic voltage waveforms $v_1(t)$, $v_2(t)$, $v_3(t)$, and $v_4(t)$:

$$\int \sin(\omega_n t)\, dt = -\frac{\cos(\omega_n t)}{\omega_n}$$

$$\int \cos(\omega_n t)\, dt = \frac{\sin(\omega_n t)}{\omega_n}$$

$$\int t \sin(\omega_n t)\, dt = \frac{\sin(\omega_n t)}{\omega_n^2} - \frac{t \cos(\omega_n t)}{\omega_n}$$

$$\int t \cos(\omega_n t)\, dt = \frac{\cos(\omega_n t)}{\omega_n^2} + \frac{t \sin(\omega_n t)}{\omega_n}$$

$$\int e^{-t/\tau} \sin(\omega_n t)\, dt = \frac{-e^{-t/\tau}[\omega_n \tau^2 \cos(\omega_n t) + \tau \sin(\omega_n t)]}{1 + (\omega_n \tau)^2}$$

$$\int e^{-t/\tau} \cos(\omega_n t)\, dt = \frac{e^{-t/\tau}[\omega_n \tau^2 \sin(\omega_n t) - \tau \cos(\omega_n t)]}{1 + (\omega_n \tau)^2}$$

The equal expressions $\omega_n = 2\pi f_n = 2\pi n f = 2\pi n/T$ will be substituted for one another whenever it is appropriate.

B.2   THE BASIC VOLTAGE WAVEFORM $v_1(t)$

The basic voltage $v_1(t)$ is defined for one full period, as follows:

$$v_1(t) = \frac{V_p}{t_r} t \quad \text{for} \quad 0 \le t \le t_r$$

$$v_1(t) = V_p \quad \text{for} \quad t_r \le t \le t_d$$

$$v_1(t) = 0 \quad \text{for} \quad t_d < t \le T$$

Therefore,

$$a_{1n} = \frac{2}{T} \int_0^T v_1(t) \cos(\omega_n t)\, dt$$

$$= \frac{2}{T} \frac{V_p}{t_r} \int_0^{t_r} t \cos(\omega_n t)\, dt + \frac{2V_p}{T} \int_{t_r}^{t_d} \cos(\omega_n)\, dt$$

$$= \frac{2}{T} \frac{V_p}{t_r} \left[\frac{\cos(\omega_n t)}{\omega_n^2} + \frac{t \sin(\omega_n t)}{\omega_n} \right]_0^{t_r} + \frac{2V_p}{T} \left[\frac{\sin(\omega_n t)}{\omega_n} \right]_{t_r}^{t_d}$$

$$= \frac{2}{T} \frac{V_p}{t_r} \left[\frac{\cos(\omega_n t_r) - 1}{\omega_n^2} \right] + \frac{2V_p}{T} \frac{\sin(\omega_n t_d)}{\omega_n}$$

$$= -\frac{4}{T} \frac{V_p}{t_r} \left[\frac{\sin^2(\omega_n t_r/2)}{(2\pi n/T)^2} \right] + \frac{V_p}{n\pi} \sin(\omega_n t_d)$$

$$= -\frac{V_p T}{n^2 \pi^2} \frac{\sin^2(\pi f_n t_r)}{t_r} + \frac{V_p}{n\pi} \sin(\omega_n t_d)$$

This follows from trigonometric identity A-13 of Appendix A, in the form

$$\cos(\omega_n t_r) - 1 = -2\sin^2\left(\frac{\omega_n t_r}{2} \right)$$

$$b_{1n} = \frac{2}{T} \int_0^T v_1(t) \sin(\omega_n t)\, dt$$

$$= \frac{2}{T} \frac{V_p}{t_r} \int_0^{t_r} t \sin(\omega_n t)\, dt + \frac{2V_p}{T} \int_{t_r}^{t_d} \sin(\omega_n)\, dt$$

$$= \frac{2}{T} \frac{V_p}{t_r} \left[\frac{\sin(\omega_n t)}{\omega_n^2} - \frac{t \cos(\omega_n t)}{\omega_n} \right]_0^{t_r} - \frac{2V_p}{T} \left[\frac{\cos(\omega_n t)}{\omega_n} \right]_{t_r}^{t_d}$$

$$= \frac{2}{T} \frac{V_p}{t_r} \frac{\sin(\omega_n t_r)}{\omega_n^2} - \frac{2V_p}{T} \frac{\cos(\omega_n t_d)}{\omega_n}$$

$$= \frac{4}{T} \frac{V_p}{t_r} \frac{\sin(\omega_n t_r/2)\cos(\omega_n t_r/2)}{(2\pi n/T)^2} - \frac{V_p}{n\pi} \cos(\omega_n t_d)$$

$$= \frac{V_p T}{n^2 \pi^2} \frac{\sin(\pi f_n t_r)\cos(\pi f_n t_r)}{t_r} - \frac{V_p}{n\pi} \cos(\omega_n t_d)$$

This follows from trigonometric identity A-9 of Appendix A, in the form

$$\sin(\omega_n t_r) = 2 \sin\left(\frac{\omega_n t_r}{2}\right) \cos\left(\frac{\omega_n t_r}{2}\right)$$

B.3 THE BASIC VOLTAGE WAVEFORM $v_2(t)$

The basic voltage waveform $v_2(t)$ is defined for one full period, as follows:

$$v_2(t) = V_p(1 - e^{-2et/t_r}) \qquad \text{for} \quad 0 \le t \le t_d$$
$$v_2(t) = 0 \qquad \text{for} \quad t_d < t \le T$$

Therefore,

$$a_{2n} = \frac{2}{T} \int_0^T v_2(t) \cos(\omega_n t)\, dt$$

$$= \frac{2V_p}{T} \int_0^{t_d} (1 - e^{-2et/t_r}) \cos(\omega_n t)\, dt$$

$$= \frac{2V_p}{T} \left[\frac{\sin(\omega_n t)}{\omega_n} \right]_0^{t_d}$$

$$- \frac{2V_p}{T} \cdot \left[\frac{e^{-2et/t_r}[\omega_n(t_r/2e)^2 \sin(\omega_n t) - (t_r/2e)\cos(\omega_n t)]}{1 + (\omega_n t_r/2e)^2} \right]_0^{t_d}$$

$$= \frac{2V_p}{T} \frac{\sin(\omega_n t_d)}{\omega_n} - \frac{2V_p}{T} \left[\frac{t_r/2e}{1 + (\omega_n t_r/2e)^2} \right]$$

$$= \frac{V_p}{n\pi} \sin(\omega_n t_d) - \frac{V_p}{n^2\pi^2} \left(\frac{eT}{t_r} \right) \frac{(\omega_n t_r/2e)^2}{1 + (\omega_n t_r/2e)^2}$$

$$= \frac{V_p}{n\pi} \sin(\omega_n t_d) - \frac{V_p T}{n^2\pi^2} \frac{\sin^2(\alpha_n)}{t_R}$$

where

$$t_R = \frac{t_r}{e} \quad \text{and} \quad \sin(\alpha_n) = \frac{\omega_n t_r/2e}{\sqrt{1 + (\omega_n t_r/2e)^2}} = \frac{\pi f_n t_R}{\sqrt{1 + (\pi f_n t_R)^2}}$$

$$b_{2n} = \frac{2}{T} \int_0^T v_2(t) \sin(\omega_n t)\, dt = \frac{2V_p}{T} \int_0^{t_d} (1 - e^{-2et/t_r}) \sin(\omega_n t)\, dt$$

$$= -\frac{2V_p}{T} \left[\frac{\cos(\omega_n t)}{\omega_n} \right]_0^{t_d} + \frac{2V_p}{T} \left[\frac{e^{-2et/t_r}[\omega_n(t_r/2e)^2 \cos(\omega_n t) + (t_r/2e)\sin(\omega_n t)]}{1 + (\omega_n t_r/2e)^2} \right]_0^{t_r}$$

$$= -\frac{2V_p}{T} \left[\frac{\cos(\omega_n t_d) - 1}{\omega_n} \right] - \frac{2V_p}{T} \left[\frac{\omega_n(t_r/2e)^2}{1 + (\omega_n t_r/2e)^2} \right]$$

$$= -\frac{2V_p}{T} \frac{\cos(\omega_n t_d)}{\omega_n} + \frac{2V_p}{T} \left[\frac{1}{\omega_n} - \frac{\omega_n(t_r/2e)^2}{1 + (\omega_n t_r/2e)^2} \right]$$

$$= -\frac{V_p}{n\pi} \cos(\omega_n t_d) + \frac{V_p}{n\pi} \left[\frac{1}{1 + (\omega_n t_r/2e)^2} \right]$$

$$= -\frac{V_p}{n\pi} \cos(\omega_n t_d) + \frac{V_p}{n\pi} \left(\frac{eT}{n\pi t_r} \right) \frac{\omega_n t_r/2e}{\sqrt{1 + (\omega_n t_r/2e)^2}} \frac{1}{\sqrt{1 + (\omega_n t_r/2e)^2}}$$

$$= -\frac{V_p}{n\pi} \cos(\omega_n t_d) + \frac{V_p T}{n^2\pi^2} \frac{\sin(\alpha_n)\cos(\alpha_n)}{t_R}$$

The final expression for b_{2n} results from the above definition of $\sin(\alpha_n)$, from which it follows that

$$\cos(\alpha_n) = \frac{1}{\sqrt{1 + (\omega_n t_r/2e)^2}} = \frac{1}{\sqrt{1 + (\pi f_n t_R)^2}}$$

B.4   THE BASIC VOLTAGE WAVEFORM $v_3(t)$

The basic voltage waveform $v_3(t)$ is defined for one full period, as follows:

$$v_3(t) = 0 \qquad \text{for} \quad 0 \le t < t_d$$

$$v_3(t) = \frac{V_p}{t_f}(t_d + t_f - t) \qquad \text{for} \quad t_d \le t \le t_d + t_f$$

$$v_3(t) = 0 \qquad \text{for} \quad t_d + t_f \le t \le T$$

Therefore,

$$
\begin{aligned}
a_{3n} &= \frac{2}{T} \int_0^T v_3(t) \cos(\omega_n t)\, dt \\[2mm]
&= \frac{2}{T}\frac{V_p}{t_f}(t_d + t_f)\int_{t_d}^{t_d+t_f} \cos(\omega_n t)\, dt - \frac{2}{T}\frac{V_p}{t_f}\int_{t_d}^{t_d+t_f} t\cos(\omega_n t)\, dt \\[2mm]
&= \frac{2}{T}\frac{V_p}{t_f}(t_d + t_f)\left[\frac{\sin(\omega_n t)}{\omega_n}\right]_{t_d}^{t_d+t_f} - \frac{2}{T}\frac{V_p}{t_F}\left[\frac{\cos(\omega_n t)}{\omega_n^2} + \frac{t\sin(\omega_n t)}{\omega_n}\right]_{t_d}^{t_d+t_f} \\[2mm]
&= -\frac{2V_p}{T}\frac{\sin(\omega_n t_d)}{\omega_n} - \frac{2}{T}\frac{V_p}{t_f}\left[\frac{\cos(\omega_n t_d + \omega_n t_f) - \cos(\omega_n t_d)}{\omega_n^2}\right] \\[2mm]
&= -\frac{V_p}{n\pi}\sin(\omega_n t_d) + \frac{4V_p}{Tt_f}\frac{\sin(\omega_n t_f/2)\sin(\omega_n t_d + \omega_n t_f/2)}{(2\pi n/T)^2} \\[2mm]
&= -\frac{V_p}{n\pi}\sin(\omega_n t_d) + \frac{V_p T}{n^2\pi^2}\frac{\sin(\pi f_n t_f)\sin(\omega_n t_d + \pi f_n t_f)}{t_f}
\end{aligned}
$$

The next to the last expression for a_{3n} follows from trigonometric identity A-8 of Appendix A, and the relationships

$$
\begin{aligned}
\cos(\omega_n t_d + \omega_n t_f) - \cos(\omega_n t_d) &= \cos\left[\left(\omega_n t_d + \frac{\omega_n t_f}{2}\right) + \frac{\omega_n t_f}{2}\right] \\
&\quad - \cos\left[\left(\omega_n t_d + \frac{\omega_n t_f}{2}\right) - \frac{\omega_n t_f}{2}\right] \\
&= -2\sin\left(\frac{\omega_n t_f}{2}\right)\sin\left(\omega_n t_d + \frac{\omega_n t_f}{2}\right)
\end{aligned}
$$

$$b_{3n} = \frac{2}{T} \int_0^T v_3(t) \sin(\omega_n t)\, dt$$

$$= \frac{2}{T} \frac{V_p}{t_f} (t_d + t_f) \int_{t_d}^{t_d + t_f} \sin(\omega_n t)\, dt - \frac{2}{T} \frac{V_p}{t_f} \int_{t_d}^{t_d + t_f} t \sin(\omega_n t)\, dt$$

$$= -\frac{2}{T} \frac{V_p}{t_f} (t_d + t_f) \left[\frac{\cos(\omega_n t)}{\omega_n} \right]_{t_d}^{t_d + t_f} - \frac{2}{T} \frac{V_p}{t_f} \left[\frac{\sin(\omega_n t)}{\omega_n^2} - \frac{t \cos(\omega_n t)}{\omega_n} \right]_{t_d}^{t_d + t_f}$$

$$= \frac{2V_p}{T} \frac{\cos(\omega_n t_d)}{\omega_n} - \frac{2}{T} \frac{V_p}{t_f} \left[\frac{\sin(\omega_n t_d + \omega_n t_f) - \sin(\omega_n t_d)}{\omega_n^2} \right]$$

$$= \frac{V_p}{n\pi} \cos(\omega_n t_d) - \frac{4V_p}{T t_f} \frac{\sin(\omega_n t_f/2) \cos(\omega_n t_d + \omega_n t_f/2)}{(2\pi n/T)^2}$$

$$= \frac{V_p}{n\pi} \cos(\omega_n t_d) - \frac{V_p T}{n^2 \pi^2} \frac{\sin(\pi f_n t_f) \cos(\omega_n t_d + \pi f_n t_f)}{t_f}$$

The next to the last expression for b_{3n} follows from trigonometric identity A-6 of Appendix A and the relationships

$$\sin(\omega_n t_d + \omega_n t_f) - \sin(\omega_n t_d) = \sin\left[\left(\omega_n t_d + \frac{\omega_n t_f}{2}\right) + \frac{\omega_n t_f}{2}\right]$$

$$- \sin\left[\left(\omega_n t_d + \frac{\omega_n t_f}{2}\right) - \frac{\omega_n t_f}{2}\right]$$

$$= 2 \sin\left(\frac{\omega_n t_f}{2}\right) \cos\left(\omega_n t_d + \frac{\omega_n t_f}{2}\right)$$

B.5 THE BASIC VOLTAGE WAVEFORM $v_4(t)$

The basic voltage waveform $v_4(t)$ is defined for one full period as follows:

$$v_4(t) = 0 \qquad\qquad \text{for} \quad 0 \le t < t_d$$
$$v_4(t) = V_p e^{-2e(t - t_d)/t_f} \qquad \text{for} \quad t_d \le t \le T$$

Therefore,

$$a_{4n} = \frac{2}{T} \int_0^T v_4(t) \cos(\omega_n t)\, dt$$

$$= \frac{2V_p}{T} \int_{t_d}^T e^{-2e(t-t_d)/t_f} \cos(\omega_n t)\, dt$$

$$= \frac{2V_p}{T} e^{2et_d/t_f} \left[\frac{e^{-2et/t_f}[\omega_n(t_f/2e)^2 \sin(\omega_n t) - (t_f/2e)\cos(\omega_n t)]}{1 + (\omega_n t_f/2e)^2} \right]_{t_d}^T$$

$$= -\frac{2V_p}{T} \frac{[\omega_n(t_f/2e)^2 \sin(\omega_n t_d) - (t_f/2e)\cos(\omega_n t_d)]}{1 + (\omega_n t_f/2e)^2}$$

$$= -\frac{V_p}{n\pi} [1 - \cos^2(\beta_n)] \sin(\omega_n t_d) + \frac{V_p}{n\pi} \sin(\beta_n)\cos(\beta_n)\cos(\omega_n t_d)$$

$$= -\frac{V_p}{n\pi} \sin(\omega_n t_d) + \frac{V_p}{n\pi} \cos(\beta_n)[\sin(\omega_n t_d)\cos(\beta_n) + \cos(\omega_n t_d)\sin(\beta_n)]$$

$$= -\frac{V_p}{n\pi} \sin(\omega_n t_d) + \frac{V_p T}{n^2\pi^2} \frac{\sin(\beta_n)\sin(\omega_n t_d + \beta_n)}{t_F}$$

where

$$\sin(\beta_n) = \frac{\omega_n t_f/2e}{\sqrt{1 + (\omega_n t_f/2e)^2}} \qquad \text{and} \qquad \cos(\beta_n) = \frac{1}{\sqrt{1 + (\omega_n t_f/2e)^2}}$$

and the final expression for a_{4n} follows from identity A-1 of Appendix A and the relationship

$$\cos(\beta_n) = \frac{2e}{\omega_n t_f} \quad \sin(\beta_n) = \frac{\sin(\beta_n)}{n\pi t_F} \qquad \text{where} \qquad t_F = \frac{t_f}{e}$$

$$b_{4n} = \frac{2}{T} \int_0^T v_4(t) \sin(\omega_n t)\, dt$$

$$= \frac{2V_p}{T} \int_{t_d}^{t_d + t_f} e^{-2e(t-t_d)/t_f} \sin(\omega_n t)\, dt$$

$$= -\frac{2V_p}{T} e^{2et_d/t_f} \left[\frac{e^{-2et/t_f}[\omega_n(t_f/2e)^2 \cos(\omega_n t) + (t_f/2e)\sin(\omega_n t)]}{1 + (\omega_n t_f/2e)^2} \right]_{t_d}^{t_d + t_f}$$

$$= \frac{2V_p}{T} \left[\frac{\omega_n(t_f/2e)^2 \cos(\omega_n t_d) + (t_f/2e)\sin(\omega_n t_d)}{1 + (\omega_n t_f/2e)^2} \right]$$

$$= \frac{V_p}{n\pi} [1 - \cos^2(\beta_n)]\cos(\omega_n t_d) + \frac{V_p}{n\pi} \sin(\beta_n)\cos(\beta_n)\sin(\omega_n t_d)$$

$$= \frac{V_p}{n\pi} \cos(\omega_n t_d) - \frac{V_p}{n\pi} \cos(\beta_n)[\cos(\beta_n)\cos(\omega_n t_d) - \sin(\beta_n)\sin(\omega_n t_d)]$$

$$= \frac{V_p}{n\pi} \cos(\omega_n t_d) - \frac{V_p T}{n^2\pi^2} \frac{\sin(\beta_n)\cos(\omega_n t_d + \beta_n)}{t_F}$$

The final expression for b_{4n} follows from the previously noted relationship between $\sin(\beta_n)$ and $\cos(\beta_n)$ and trigonometric identity A-3 of Appendix A.

BIBLIOGRAPHY

Antennas

Balanis, C. A., *Antenna Theory: Analysis and Design*. Harper & Row, New York, 1982.

Collin, R. E., *Antennas and Radiowave Propagation*. McGraw–Hill, New York, 1985.

Johnson, R. C., and Jasik, H., *Antenna Engineering Handbook*, 2d ed. McGraw–Hill, New York, 1984.

Kraus, J. D., *Antennas*. McGraw–Hill, New York, 1950.

Kraus, J. D., *Antennas*, 2d ed. McGraw–Hill, New York, 1988.

Lee, K. F., *Principles of Antenna Theory*. Wiley, New York, 1984.

Schelkunoff, S. A., and Friis, H. T., *Antennas—Theory and Practice*. Wiley, New York, 1952.

Silver, S. (Ed.), *Microwave Antenna Theory and Design*. McGraw–Hill, New York, 1949.

Stutzman, W. L., and Thiele, G. A., *Antenna Theory and Design*. Wiley, New York, 1981.

Weeks, W. L., *Antenna Engineering*. McGraw–Hill, New York, 1968.

Electrical Engineering

Jay, F. (Ed.), *IEEE Standard Dictionary of Electrical and Electronics Terms*. The Institute of Electrical and Electronics Engineers, New York, 1984.

Simonyi, K., *Foundations of Electrical Engineering*. Macmillan, New York, 1963.

Electromagnetic Compatibility

Ficchi, R. F., *Electromagnetic Compatibility*. Hayden, New York, 1971.

Ott, H. W., *Noise Reduction Techniques in Electronic Systems*. Wiley, New York, 1988.

Electromagnetic Theory

Adler, R. B., Chu, L. J., and Fano, R. M., *Electromagnetic Energy Transmission and Radiation.* Wiley, New York, 1960.

Hayt, W. H., Jr., *Engineering Electromagnetics.* McGraw–Hill, New York, 1981.

Holt, C. A., *Introduction to Electromagnetic Fields and Waves.* Wiley, New York, 1963.

Jordan, E. C., and Balmain, K. G., *Electromagnetic Waves and Radiating Systems,* 2d ed. Prentice–Hall, Englewood Clifs, NJ, 1968.

Kraus, J. D., and Carver, K. R., *Electromagnetics,* 2d ed. McGraw–Hill, New York, 1973.

Plonsey, R., and Collin, R. E., *Principles and Applications of Electromagnetic Fields.* McGraw–Hill, New York, 1961.

Ramo, S., Whinnery, J. R., and Van Duzer, T., *Fields and Waves in Communication Electronics.* Wiley, New York, 1965.

Rojansky, V., *Electromagnetic Fields and Waves.* Dover, New York, 1979.

Stratton, J. A., *Electromagnetic Theory.* McGraw–Hill, New York, 1941.

Electronics

Gray, P. E., and Searle, C. L., *Electronic Principles.* Wiley, New York, 1969.

Terman, F. E., *Electronic and Radio Engineering,* 4th ed. McGraw–Hill, New York, 1955.

Mathematics

Davis, H. F., *Fourier Series and Orthogonal Functions.* Allyn & Bacon, Boston, 1963.

Gellert, W., Kustner, H., Hellwich, M., and Kastner, H. (Eds.), *The VNR Concise Encyclopedia of Mathematics.* Van Nostrand Reinhold, New York, 1977.

Hodgman, C. D. (Ed.), *Mathematical Tables from Handbook of Chemistry and Physics,* 10th ed. Chemical Rubber, Cleveland, OH, 1954.

Tolstov, G. P. (Translated by Silverman, R. A.), *Fourier Series.* Prentice–Hall, Englewood Cliffs, NJ, 1962.

Measurements

Klein, H. A., *The World of Measurements.* Simon & Schuster, New York, 1974.

Terman, F. E., and Pettit, J. M., *Electronic Measurements,* 2d ed. McGraw–Hill, New York, 1952.

Physics

Eisberg, R. M., and Lerner, L. S., *Physics: Foundations and Applications.* McGraw–Hill, New York, 1981.

Sears, F. W., and Zemansky, M. W., *University Physics.* Addison–Wesley, Reading, MA, 1955.

INDEX